HISTORICAL RECORDS

OF THE

FORTY-FOURTH.

COLOURS OF THE 44TH REGIMENT, CARRIED DURING THE CRIMEAN WAR.
PRESENTED IN 1843. PLACED IN ST PETER'S CHURCH, COLCHESTER, 1857.

HISTORICAL RECORD

OF THE

FORTY-FOURTH,

OR THE

EAST ESSEX REGIMENT.

COMPILED BY

THOMAS CARTER,

Adjutant-General's Office,

Author of "Curiosities of War" and "Medals of the British Army."

SECOND EDITION.

"'Twere a concealment,
Worse than a theft, no less than a traducement,
To hide your doings."—*Shakespeare.*

WITH ILLUSTRATIONS.

THE FORTY-FOURTH

BEARS ON THE REGIMENTAL COLOUR

"THE SPHINX," "EGYPT."

"BADAJOZ," "SALAMANCA,"

"PENINSULA,"

"BLADENSBURG,"

"WATERLOO,"

"AVA,"

"ALMA," "INKERMAN,"

"SEVASTOPOL,"

"TAKU FORTS."

PREFACE TO FIRST EDITION.

THE official publication of the Historical Records of the Regiments of the British Army was authorised by the following General Order, dated Horse Guards, 1st January, 1836 :—

HIS MAJESTY has been pleased to command that, with the view of doing the fullest justice to Regiments, as well as to Individuals who have distinguished themselves by their Bravery in Action with the Enemy, an Account of the Services of every Regiment in the British Army, shall be published under the superintendence and direction of the Adjutant-General; and that this Account shall contain the following particulars, viz. :—

—— The Period and Circumstances of the Original Formation of the Regiment; The Stations at which it has been from time to time employed; The Battles, Sieges, and other Military Operations in which it has been engaged, particularly specifying any Achievement it may have performed, and the Colours, Trophies, &c., it may have captured from the Enemy.

—— The Names of the Officers, and the Number of Non-Commissioned Officers and Privates Killed or Wounded by the Enemy, specifying the Place and Date of the Action.

—— The Names of those Officers who, in consideration of their Gallant Services and Meritorious Conduct in Engagements with the Enemy, have been distinguished with Titles, Medals, or other Marks of His Majesty's gracious favour.

—— The Names of all such Officers, Non-Commissioned Officers, and Privates, as may have specially signalized themselves in Action.

And,

—— The Badges and Devices which the Regiment may have been permitted to bear, and the Causes on account of which such Badges or Devices, or any other Marks of Distinction, have been granted.

By Command of the Right Honourable

GENERAL LORD HILL,

Commanding-in-Chief.

JOHN MACDONALD,

Adjutant-General.

Several regimental histories were in consequence compiled by Mr. Richard Cannon, Principal Clerk of the Adjutant-General's Office, and published under the patronage of his late Majesty WILLIAM THE FOURTH, and of Her Majesty THE QUEEN. Upon Mr. Cannon's retirement, Lieut.-General Sir George Wetherall, then Adjutant-General, in consequence of my having assisted that gentleman in preparing these Records, recommended me to the Lords Commissioners of the Treasury as the future Editor, but it was decided that the work should not be continued.

This will account for the History of the 44th not having been already published by authority; in consequence, however, of all its Records having been lost at Cabul, and with the view of carrying out the General Order before quoted, application was made to me by Colonel Staveley to trace the services of the Regiment from the period of formation in 1741.

The Record has now been completed, and with that natural feeling of *esprit-de-corps,* which the publication of these several Histories by authority was specially intended to encourage, these pages are now submitted to the Army and to the Public, in the hope that they will not prove altogether unacceptable; while at the same time, they are designed to meet the desire of those who have served, are serving, or who have had relatives, in the "TWO FOURS."

To Colonel Burney, K.H., and Lieut.-Colonel Pearce, K.H., the Regiment is greatly indebted for much valuable and interesting information, which, without their aid, could not have been obtained.

T. C.
D. B.

HORSE GUARDS,
31st December, 1864.

CONTENTS.

Year.		Page.
1741	Formation of the 55th Regiment, afterwards numbered the 44th	2
——	Colonel James Long appointed Colonel of the Regiment	—
——	Names of Officers	3
1742	The Regiment stationed in Great Britain	4
1743	Lieutenant-Colonel John Lee appointed Colonel of the Regiment	—
1744	War with France	—
1745	Landing in Scotland of Prince Charles Edward	—
——	Battle of Preston Pans	6
1748	Termination of the War	7
——	The 55th numbered the 44th Regiment	—
1749	Removed to Ireland	—
1751	Colonel Sir Peter Halkett, Bart., appointed Colonel of the Regiment	—
——	Royal Warrant regulating the clothing, standards, facings, number, and rank of Regiments	8
1755	Embarked for North America	—
——	Expedition against Fort du Quesne, under Major-General Braddock	—
——	Action near Fort du Quesne	9
——	Colonel Robert Ellison appointed Colonel of the Regiment	13
1756	Major-General James Abercromby appointed Colonel of the Regiment	—
——	Commencement of the Seven Years' War	—
1758	Attack of Fort Ticonderoga	14
1759	Operations against Fort Niagara	16
——	Capture of the Fort	17
1760	Surrender of Fort Levi	18

Year.		Page.
1760	Conquest of Canada completed	19
1765	Regiment returned home	—
1766	Stationed in Ireland	—
1775	War with the American Colonists	20
—	Regiment Embarked for Boston	—
1776	Expedition against Staten Island	—
—	Landing on Long Island	21
—	Action at Brooklyn	—
—	Capture of New York	—
—	Action at White Plains	—
—	Reduction of Forts Washington and Lee	—
1777	Operations against Danbury	22
—	Action at Ridgefield	—
—	Affair at Compo	23
—	Expedition against Philadelphia	—
—	Battle of Brandywine	—
—	Surprise of General Wayne	24
—	Battle of Germantown	—
—	Occupation of Philadelphia	25
1778	Expedition against Bedford	—
1779	Regiment Stationed near New York	26
1780	Embarked for Canada	—
1781	Major-General Rainsford appointed Colonel of the Regiment	—
1782	Styled, by Royal Authority, the "East Essex Regiment."	27
1783	Conclusion of Peace with the United States	—
1786	Regiment Returned to England from Canada	—
1789	Removed to Scotland	—
1792	Embarked for Ireland	—
1793	Flank Companies Embarked for Barbadoes	28
1794	Expedition against Martinique	—
—	Capture of St. Lucia and Guadaloupe	—
—	Battalion Companies Embarked for Ostend	29
—	Incorporated with the Army of the Netherlands, under the Duke of York	—

Year.		*Page.*
1794	Attack on Boxtel	29
——	Services of Flank Companies at Guadaloupe	30
1795	British Troops Removed from Holland to Germany	31
——	Return of the Regiment to England	—
——	Embarked for the West Indies	—
1796	Recapture of St. Lucia	32
——	Employed against the Caribs	33
1797	Casualties amongst the Officers in the West Indies	—
——	Return of the Regiment to England	—
1798	Embarked for Gibraltar	34
——	Services of Major Donkin, with the Expedition under Major-General Coote	—
1800	The Regiment selected for the Expedition to Egypt	—
1801	Landing near Alexandria	—
——	Action of the 13th of March	35
——	Battle of Alexandria	—
——	Death of Sir Ralph Abercromby	36
——	Capture of Rosetta	—
——	Siege of Cairo	37
——	Siege of Alexandria	—
——	Surrender of the Garrison	—
——	Honours Conferred for the Egyptian Campaign	—
——	The SPHINX, and the word "EGYPT," authorised for the Regimental Colour	—
——	The Regiment Embarked for Malta	—
——	Returned to England	—
1802	Proceeded to Ireland	38
1803	Renewal of War	—
——	Formation of the Second Battalion	—
——	The Regiment Removed to England	—
1805	Embarked on Secret Service, and afterwards Stationed at Malta	—
1808	Flank Companies Employed in Sicily	39
——	Services of Major Powlett at the Battle of Maida	—
——	Flank Companies Rejoin at Malta	—

Year.		*Page.*
1808	The Regiment Embarked for Sicily	39
—	Reminiscences of Services in that Island	—
1809	General Sir Thomas Trigge, K.B., appointed Colonel of the Regiment	—
—	Employed, under Lieutenant-General Sir John Stuart, in the Expedition against Naples	—
—	Capture of the Islands of Ischia and Procida	40
—	Description of the Uniform of the Period	—
—	Attempt on the Castle of Scylla	41
—	Expedition against the Ionian Islands	—
1810	Stationed in Sicily	—
—	Landing of General Murat in that Island	—
1811	Embarkation of the Regiment for Malta	—
—	Noble Conduct of Lieutenant Hill	—
1812	Flank Companies Embarked for Sicily	42
—	Return of—to Malta	—
1813	The Regiment Proceeded to Sicily	—
—	Embarked for Minorca	—
—	———— the East Coast of Spain	—
—	Joined the Army before Tarragona	—
1814	March to Villafranca	—
—	Termination of Hostilities	—
—	The word "PENINSULA" authorised for the Colours	—
—	General the Earl of Suffolk appointed Colonel of the Regiment	43
—	War with the United States	—
—	The Regiment Embarked for North America	—
—	Landing near St. Benedict	44
—	Advance on Washington	45
—	Battle of Bladensburg	—
—	The word "Bladensburg" authorised for the Regimental Colour	46
—	Action near Baltimore	47
—	Landing at the American Camp near the Potomac	49
—	Embarkation of the Troops for the West Indies	—

Year.		*Page.*
1814	Expedition against New Orleans	50
——	Landing near that Place	—
——	The Americans Repulsed	—
1815	Attack on the Enemy's Lines	51
——	The Troops withdrawn from before New Orleans	52
——	Casualties of the Regiment	—
——	Operations against Mobile	—
——	Surrender of the Garrison	—
——	Termination of Hostilities	53
——	Disembarkation of the Regiment at Cork	—
——	Court-Martial on Lieutenant-Colonel the Hon. Thomas Mullins	—

SERVICES OF THE SECOND BATTALION.

1803	Establishment of the Battalion	54
——	Names of Officers	—
1804	Removed from Mallow to the Isle of Wight	55
——	Embarked for Guernsey	—
1807	The Scotch Snuff-mull presented by Captain Fyffe	—
1809	Removed to Alderney	—
1810	Embarked for Cadiz	—
——	Defence of Matagorda	—
——	Proceeded to Portugal	56
——	Joined the Army at Torres Vedras	—
1811	Pursuit of Marshal Massena	57
——	Lieutenant Pearce and the Light Company	—
——	Action near Sabugal	—
——	Portugal Rescued from the French	—
——	Battle of Fuentes d'Onor	—
——	Paymaster Williams and his Accounts	58
——	Pursuit of the Garrison of Almeida	—
——	Employed in Watching Marshal Marmont's Movements.	59
1812	Siege and Capture of Ciudad Rodrigo	60
——	Siege of Badajoz	—

Year.		Page.
1812	The Colours of the Regiment the first planted on the Bastion	61
——	Services of Captain Guthrie and Lieutenant Pearce	—
——	Anecdote of Bugler Muchian	—
——	Casualties amongst the Officers	62
——	The word "BADAJOZ" authorised for the Colour	63
——	Advance towards Salamanca	—
——	Capture of the Forts	—
——	Race of the Hostile Columns at the Guarena	64
——	Battle of Salamanca	—
——	Capture of the Eagle by Lieutenant Pearce	65
——	Services of that Officer	—
——	French drum taken by the Regiment	—
——	The word "SALAMANCA" authorised for the Colour	67
——	Siege of Burgos	68
——	Action at Villa Muriel	—
——	Withdrawal from that Place	—
——	Lieutenant Grier's Greyhounds and the Hare	—
——	Casualties of the Regiment	—
——	Sergeant Farrell the Representative of his Company	69
——	Sufferings on the Retreat from Burgos	—
——	The Battalion Formed into Four Companies	70
1813	Embarkation of the Remains of the Six Companies for England	—
——	Incorporation of the Four Companies left in the Peninsula with the Four Companies of the Second Battalion of the 30th Regiment, and numbered the Provisional Battalion	—
——	Sobriquet of the "Little Fighting Fours"	—
——	Provisional Battalion Embarked for England	—
——	The word "PENINSULA" authorised for the Colour	—
——	Furnished the Guard for the Prince Regent at Brighton	71
——	Embarked for Holland, under Lieutenant-General Sir Thomas Graham	—
1814	Advance on Antwerp	—
——	The two Actions at Merxem	—

Year.		*Page.*
1814	Attack on Bergen-op-Zoom	71
——	Lieutenant-Colonel Carleton and the Cossacks	72
——	Details of the Assault	—
——	Services of Captain Burney and other Officers	73
——	Severe Casualties of the Battalion	74
——	Gallantry of Private McCullop	—
——	Conclusion of Peace	75
1815	Stationed at Ostend	—
——	Return of Napoleon to France	76
——	Opening of the Campaign	—
——	Action at Quatre Bras	77
——	Lieutenant-Colonel Hamerton and the French Cavalry	78
——	Received by the Battalion in Line	—
——	Gallantry of Ensign Christie	79
——	Preservation of the Colour	80
——	Ball afterwards given to the Officers by Lieutenant-General Hamerton	—
——	Subsequent Attack of the French Cavalry	—
——	Captain Burney's Reminiscences	82
——	Movement upon Waterloo	83
——	Battle of Waterloo	—
——	French Attack on La Haye Sainte	84
——	Death of Sir Thomas Picton	—
——	Last Effort of Napoleon	—
——	Arrival of the Prussians	—
——	The Repeater Watch taken by Ensign Dunlevie	85
——	Pensioner Thomas Brooks, a Relic of Captain Burney's Company	86
——	Casualties on the 16th and 18th of June	—
——	The "Dead Alive"	—
——	Honours for the Campaign	87
——	The word "WATERLOO" authorised for the Colour	—
——	Description of the Waterloo Medal	—
——	Names of the Officers to whom it was Awarded	—
——	Lieutenant Grier's Dog	88

Year.		*Page.*
1815	Lieutenant Campbell and his Father at Waterloo.......	88
——	March on Paris	89
——	General Order by the Duke of Wellington..................	—
1816	Return of the Battalion to England........................	90
——	Memorial of the Officers against Disbanding the Battalion	91
——	The Commander-in-Chief's Reply........................	93
——	Disbanded at Dover..	94
——	The Silver Plate converted into a handsome Soup Tureen, and Presented by the Officers of the Second Battalion to the Regiment........................	—

SERVICES OF THE REGIMENT—Continued.

1816 to 1821	Stationed in Ireland........................	95
1820	Lieutenant-General Gore Brown appointed Colonel of the Regiment	—
——	New Colours Presented	—
1822	Embarked for Calcutta	—
1823	Proceeded to Dinapore	—
——	Casualties from a Storm, and Upsetting of the Boats while on Passage up the Ganges	—
——	The Band Instruments Destroyed........................	96
——	Casualties from the Climate of India	—
1824	Removed to Calcutta	—
——	Death of Ensign Gilbert, the Father of Lola Montez ...	—
——	War against the Court of Ava	—
——	Embarked to Join the Force assembled at Chittagong...	97
1825	Advance on Rangoon	98
——	Attacks on Ramoo and Ramree........................	—
——	Advance upon Arracan	—
——	Forcing of the Padawa Pass	99
——	Attack on the Burmese Works at Mahattee	—
——	Capture of Arracan........................	100
——	Services of Lieutenant Whitney	—
——	Four Companies, under Captain Shelton, Detached to Ramree	—

Year.		*Page*
1825	Sufferings from Fever of Captain Mackrell's Company, Detached against Talak	101
——	Casualties of the Regiment from Fever	—
——	Withdrawn from Arracan	102
——	Peculiarities in the Malady	—
1826	Termination of the Burmese War	103
——	General Orders by the Governor-General	—
——	The word "AVA" authorised for the Regimental Colour.	104
——	Grant of a Clasp for the Campaign	105
——	The Regiment Proceeded to Ghazeepore	—
1828	Removed to Cawnpore	—
1833	Marched from Cawnpore to Chinsurah	—
1835	Proceeded to Calcutta	—
1837	Returned to Ghazeepore	—
——	Casualties from Cholera	—
1838	Marched to Kurnaul	—
1840	Ordered to Proceed to Cabul	—
1841	Brigadier Shelton's Campaign against the Refractory Tribes	106
——	Capture of the Forts in the Nazian Valley	—
——	Acknowledgment of the Service in Government Orders.	—
——	Arrival at Cabul	107
——	Expedition against a Refractory Tribe in the Zoormut Valley	—
——	Outbreak at Cabul	—
——	Names of Officers Present	—
——	Progress of the Insurrection	108
——	Attack on the Fort at Mahomed Shereef	109
——	Storming the Rika-bashee Fort	111
——	Death and Services of Lieutenant-Colonel Mackrell	—
——	Attack on the Heights of Be-maroo	112
——	Fort of Mahomed Shereef Captured by the Affghans ...	113
——	Gallantry of Private John Stuart	—
——	Negotiations Opened with the Chiefs	—
——	Interview with Akbar Khan, and Murder of the British Envoy (Sir William Macnaghten)	114

Year.		Page.
1842	Retreat from Cabul	115
——	First Day's Sufferings	—
——	Second Day's Sufferings	—
——	Third Day's Sufferings	116
——	Massacre in the Khoord Cabul Pass	117
——	Fourth Day's Sufferings	118
——	Communications again Opened with Akbar Khan	—
——	Desertion of the Second Shah's Cavalry	—
——	Fifth Day's Sufferings	119
——	Conflict in the Tezeen Pass	120
——	Sixth Day's Sufferings	—
——	Conference with Akbar Khan	—
——	Major-General Elphinstone and Brigadier Shelton Detained by the Sirdar	124
——	Officers Killed in Jugdulluck and in the Pass	—
——	Seventh Day's Sufferings	125
——	Eighth Day's Sufferings	—
——	Last Stand of the Regiment	126
——	Names of Officers Killed during the Retreat	127
——	Preservation of the Regimental Colour by Lieutenant Souter	128
——	March of Major-General Pollock to Jellalabad	131
——	Death of Major-General Elphinstone	—
——	Termination of the Siege of Jellalabad	—
——	Advance of the British on Cabul	—
——	Junction of Major-General Knott's Force with that under major-General Pollock at Cabul	132
——	Release of the Prisoners	—
——	Names of those Belonging to the Regiment	—
——	Return of the Troops to India	133
1843	Court-Martial on Brigadier Shelton, and Testimonies to his Gallantry	—
——	Casualties in the Detachment which had Proceeded from Cawnpore to Join the Regiment	135
——	Embarked for England	136

Year.		Page.
1843	Lieutenant-General the Hon. Patrick Stuart appointed Colonel of the Regiment	136
——	Completion of the Regiment to its Establishment	—
——	Colonel Burney's Visit to Colonel Shelton at Gosport	—
——	New Percussion Muskets Issued to the Regiment	137
——	Presentation of New Colours	—
——	The Old Colour Saved by Lieutenant Souter Deposited in Alverstoke Church	138
——	Monument Placed there to the Memory of the Officers and Men who Fell in Affghanistan	—
1845	Regiment Embarked for Ireland	139
——	Death and Services of Colonel Shelton	—
1846	Ordered to Recruit to Twelve Companies	—
1847	Completed to its Establishment	140
1848	Organised into First and Reserve Battalions	141
——	Embarked for Malta	—
——	Depôt Stationed at Parkhurst	—
1850	Regiment Reduced to Ten Companies	—
——	Reserve Battalion Broken up	—
——	Casualties from Cholera at Malta	—
1851	Removed to Gibraltar	—
1852	Death of Lieutenant and Adjutant Colpoys	—
1853	Stationed at Gibraltar	—
1854	War with Russia	142
——	The Regiment Embarked to Join the Expeditionary Army under Lord Raglan	—
——	Arrival at Malta	—
——	Partly Armed with Minié Rifles	143
——	Proceeded to Gallipoli	—
——	Formed part of the Third Division, and Moved to Varna	—
——	Portion of the Regiment, under Major Staveley, employed with the Cavalry Division	144
——	Remarkable Flight of Locusts at Varna	—
——	Completed with Minié Rifles	—
——	Embarked at Varna for the Crimea	—
——	Names of Officers	—

B 2

Year.		*Page.*
1854	Landing in the Crimea	145
——	Battle of the Alma	146
——	Part taken by the Regiment therein	147
——	Absence of the Dog "Bob" from the Scots' Fusilier Guards Accounted for	148
——	Assistant-Surgeon Thomson and Private Magrath's Devotedness to the Russian Wounded	149
——	Flank March on Sebastopol	151
——	First Bombardment of Sebastopol	152
——	Battle of Inkermann	153
——	Services of the Third Division	155
——	The Regiment Augmented to Sixteen Companies	——
——	Names of Non-commissioned Officers Appointed to Commissions	——
——	Christmas Day in the Crimea	156
——	Casualties of the Regiment	——
——	Sufferings in the Trenches	157
1855	Lieutenant-General Sir Frederick Ashworth appointed Colonel of the Regiment	——
——	Major-General Eyre's Commendation of the Regiment..	158
——	Second, Third, and Fourth Bombardments of Sebastopol	——
——	Attack on the Cemetery	——
——	Severe Casualties of the Regiment	163
——	Names of the Officers	——
——	Death of Lord Raglan	164
——	Major-General Simpson Appointed to the Command of the British Forces in the Crimea	——
——	Colonel the Hon. A. Spencer appointed to the Command of a Brigade, and Colonel Staveley to that of the Regiment	——
——	Final Bombardment of Sebastopol	165
——	Attack on the Redan	——
——	Termination of the Siege of Sebastopol	166
——	Names of Officers of the Regiment Present at its Fall...	——
——	The words "ALMA" "INKERMAN" and "SEVASTOPOL" authorised for the Regimental Colour	167

Year.		Page.
1855	Decorations, Medals, and Rewards for the Army in the Crimea	167
——	Names of Officers and Men who Received them	168
——	The Victoria Cross granted to Sergeant McWhiney	169
1856	Suspension of Hostilities and Treaty of Peace	——
——	Arrival of the Regiment in England	——
——	Inspected by Her Majesty	——
——	Reduced to Twelve Companies	170
1857	The Indian Mutiny	——
——	The Regiment Ordered for Madras	——
——	Presentation of New Colours by Major-General the Hon. Sir James Scarlett, Commanding the Troops in the South-Western District	——
——	Embarkation of the Regiment	171
——	The Depôt Moved to Colchester	——
——	Arrival of the Regiment at Madras	172
1858	Major-General Thomas Reed, C.B., appointed Colonel of the Regiment	——
1859	The Regiment Ordered to Proceed on Active Service	——
1860	Renewal of the War with China	——
——	The Regiment Embarked for China	——
——	Landing at Pehtang	173
——	Names of Officers with the Regiment	——
——	Attack of the Sinho Intrenchment	——
——	Operations of the Column under Brigadier Staveley	174
——	Capture of Tangku	175
——	Attack on the Taku Forts	176
——	Operations of the Storming Party under Lieutenant-Colonel MacMahon	——
——	The Victoria Cross granted to Lieutenant Rogers and Private M'Dougall, of the Regiment	177
——	Casualties of the Regiment	178
——	The words "TAKU FORTS" authorised for the Regimental Colour	——
——	Shanghai Threatened by the Taepings	——
——	Capture of Pekin and Termination of the War	179

Year.		*Page.*
1860	Honours for the Chinese Campaign	179
——	Colonel Staveley Appointed to the Command of the Troops in China, with the Rank of Brigadier	——
1861	The Regiment Ordered to Return to India	180
——	Brigadier-General Crawford's Complimentary Order on the Regiment leaving Hong Kong	——
——	Sir John Michel's Confirmation of the above Order	181
——	Arrival of the Regiment at Bombay	——
——	Removed to Belgaum	——
1862	Presentation of the Victoria Cross to Private M'Dougall	182
1865	Industrial Exhibition	——
——	Embarked for England	183
1866	Fire at Sea	184
——	Arrived in England	——
——	Stationed at Dover	——
1867	Proceeded to Aldershot	——
——	Snider Rifles issued	——
1868	Embarked for Ireland	——
1869	Glengarry Cap taken into wear	185
1870	Under Orders for the Cape of Good Hope	——
1871	Embarked for India	186
——	Stationed at Kampti	——
1872	Roll of Officers on 1st January	——
1873	Second Industrial Exhibition	187
——	Quartermaster McHarg embarked for England	——
1874	Sphinx and "Egypt" allowed to be worn by Officers on Forage Caps	——
1875	Proceeded to Secunderabad	——
1877	Embarked for Burmah	188
1879	Held in readiness for active service	——
1880	Left wing marched from Toungoo to Thayetmyo	——
1881	Deprived of Old Title, and designated "1st Battalion Essex Regiment"	189

APPENDIX.

	Page
Succession of Colonels of the Regiment	190
Succession of Commanding Officers of the Regiment............	191
Names of Men who Served in the Trenches before Sebastopol the Whole Time	195
Names of Men who were in the Crimea the Whole Time.........	196
Names of Men Mentioned in Regimental Orders during the Crimean War	197
Names of Men who Distinguished themselves during the Chinese Campaign of 1860..................................	—

44th. Regiment of Foot 1742.

HISTORICAL RECORD

OF THE

FORTY-FOURTH,

OR THE

EAST ESSEX REGIMENT;

ORIGINALLY NUMBERED THE

FIFTY-FIFTH REGIMENT.

In consequence of the aggressions committed on the 1739
commerce of Great Britain in the West Indies and South America by Spain, war was proclaimed against that power on the 23rd of October, 1739. Augmentations were accordingly made in the army and navy, ten regiments of Marines being raised in this and the following year, which were numbered from the 44th to the 53rd Regiment.

The above-mentioned 44th (Wolfe's) Regiment, as will be afterwards shown, is not the present 44th Regiment of the Line.

While the war was being carried on between Great 1740
Britain and Spain, Charles VI., Emperor of Germany, died, and the succession of his daughter, the Archduchess Maria Theresa, being disputed by the Electors of Bavaria and Saxony, also by the Kings of Prussia and Spain, a continental war was the result, in which England and France, acting in the first instance as auxiliaries, finally became principals in the contest, which is known as the "*War of the Austrian Succession.*"

1741 In January, 1741, seven additional regiments were raised, for the regular Infantry, and were numbered in succession to the ten regiments of Marines, from the 54th to the 60th Regiment, as follows :—

54th	Regt.,	comd. by Col.	Thomas Fowke,	now the	43rd	Regt
55th	,,	,,	James Long,	,,	44th	,,
56th	,,	,,	D. Houghton,	,,	45th	,,
57th	,,	,,	John Price,	,,	46th	,,
58th	,,	,,	J. Mordaunt,	,,	47th	,,
59th	,,	,,	J. Cholmondeley	,,	48th	,,
60th	,,	,,	H. de Grangue, disbanded in 1748.			

The Regiment, which is the subject of this record, was therefore known as the 55th from its formation until the Treaty of Aix-la-Chapelle, in 1748, when, in consequence of the disbandment of Col. Spotswood's (afterwards Gooche's) American Provincial Corps, then numbered the 43rd Regiment, and of the ten Marine Regiments from the 44th to the 53rd, the numerical titles of six of the seven regiments above specified were changed, and the 55th became the 44th Regiment, which has since retained that number.

Colonel James Long, from the present Grenadier Guards, was the first Colonel of the 55th Regiment, his commission being dated 7th January, 1741, and on the 17th of that month the following letter was addressed to him from the War Office :—

" *War Office, 17th January, 1741.*

" SIR,—His Majesty having thought fit to order a " regiment of foot to be forthwith raised under your " command, which is to consist of ten companies of three " serjeants, three corporals, two drummers, and seventy " effective private men in each company, besides commission " officers, and to grant a warrant for allowing two pounds " for each private man as levy-money, and to authorise the " Commissary-General of the Musters to make out Muster " Rolls *compleat* for two musters from the 25th of December " last, the commencement of your establishment, the better " to enable your officers to raise good and able men ; I am " thereupon commanded by Mr. Secretary-at-War to acquaint

"you with this, and to tell you His Majesty expects you 1741
"will take care to have your regiment *compleat* at the
"expiration of the said two musters.

"I am further to acquaint you that the proper orders "will be sent for issuing the necessary arms, as usual, out "of His Majesty's Stores of Ordnance, as also to the "Paymaster-General of the Forces, to pay you fourteen "hundred pounds, being two pounds per man, as levy money "for seven hundred effective private men, and likewise to "issue to you two months' subsistence for the whole regiment "from the 25th day of December last inclusive.

"I am, &c.,
"(Signed) RD. ARNOLD."

"Colonel Long."

The following officers were appointed to commissions in the regiment :—

Colonel and Captain of a Company—James Long.
Lieut.-Colonel and Captain of a Company—Peter Halkett.
Major and Captain of a Company—William Shewen.

Captains.

Thomas Mason.
David Braimer.
Basil Cochrane.
Russell Chapman.
Charles Tatton.
Mathew Aylmer.
Charles Knipe.

Captain-Lieutenant—Durand Therond.

Lieutenants.

David Kennedy.
Leonard Hewetson.
George Welbourne.
James Sandilands.
Jesse Shaftoe.
Nicholas Dunbar.
Samuel Rogers (*Adjt.*)
John Bickerstaff.
George Watson.
John Dale.

Ensigns.

Michael Alcock.
Thomas Faulkener.
—— Dalzell.
John Archer.
George Davies.
West Diggs.
David Drummond.
Sir John Elphinstone.
William Cunningham.

1741 *Adjutant*—Samuel Rogers.
Quarter-master—Hugh Vans.
Surgeon—William Trotter.
Chaplain—Edmund Morris.

1742 During the year 1742 the regiment was stationed in Great Britain.

1743 On the 11th March, 1743, Lieut.-Colonel John Lee, of the 4th Foot, was promoted to the Colonelcy of the 55th Regiment, in succession to Colonel Long, appointed to the 4th Marines.

1744 France and England now became principals in the contest, and on the 20th of March, 1744, the former declared war against the latter, which was met by a counter-declaration by King George II. on the 29th of the same month.

1745 The 55th continued in Great Britain during this and the following year. In July, 1745, Prince Charles Edward, eldest son of the Pretender, arrived in the Highlands of Scotland, where he was joined by several clans. The 55th Regiment at this period formed part of the force in Scotland, and Lieut.-General Sir John Cope, the Commander-in-Chief in North Britain, assembled at Stirling all the troops under his orders, which consisted of about fourteen hundred men. He subsequently advanced towards the great road named the Chain, leading through the Highlands to Inverness, and after a laborious march arrived at Dalwhinnie on the 25th of August. Here the intelligence was received that the rebels were posted at Corryarrack, seventeen miles distant, upon which Sir John Cope continued his march through Badenoch to Inverness, thus leaving the South of Scotland unprotected. The young Pretender accordingly improved this unexpected advantage, and immediately entered the district of Athol, seized the castle of Blair, proceeded afterwards to Perth and Dundee, proclaiming his father by new magistrates of his own appointment, levying the public money, and assuming other acts of royalty. The numbers of his followers had increased to four thousand men; and on the 11th of September the young Chevalier marched from Perth, passed the Forth on the 13th, and on the 16th

of that month, at night, arrived in the vicinity of Edinburgh. 1745
At five o'clock on the following morning the city was unaccountably surrendered to him without resistance. He then made his public entry, attired in Highland costume, and occupied the royal palace of Holyrood House. General Guest, who commanded the Garrison of Edinburgh Castle, removed the Bank and the effects of the principal inhabitants into that fortress, which greatly disappointed the young Prince, who had expected to gain possession of the treasure. His father was afterwards proclaimed with great ceremony at the High Cross, as King of Great Britain and Ireland.

Lieut.-General Sir John Cope, in the meanwhile, had marched with his troops from Inverness to Aberdeen, where they took shipping, and landed at Dunbar, twenty-seven miles east of Edinburgh, on the 18th of September. He was here reinforced by Brigadier-General Fowke, with two regiments of Dragoons, from Edinburgh, and the next day advanced towards that city to observe the disposition of the rebels, who were now increased to upwards of five thousand men.

On the 20th September Sir John Cope encamped in the neighbourhood of *Preston Pans*, near the sea, and seven miles from Edinburgh; his army consisted of the following regiments:—

Gardiner's (13th) and Hamilton's (14th) Dragoons	567
Two Companies of Guise's (6th) and eight of Lascelles' (47th) Foot	570
Five Companies of Lee's (afterwards 44th) Regiment	291
Murray's (46th) Regiment	580
Highlanders	183
Total	2,191

Information being received of the approach of the enemy, Sir John Cope drew up his army at *Gladsmuir Heath*, between the hamlets of *Preston Pans* and Cockenzie. About three o'clock on the morning of the 21st of September, the insurgent Highlanders commenced their march, and before

1745 daybreak a chosen band of these hardy mountaineers advanced with great celerity and intrepidity to attack the Royalists. As they drew near, they raised a fearful yell, fired a volley, threw down their muskets, and rushed sword in hand upon the troops which guarded the artillery. The sudden advance of the Highlanders in the dark, their superior numbers, and peculiar mode of fighting, dismayed the two hundred soldiers appointed to guard the artillery on the right, who saw themselves assaulted by more than three times their own numbers, and as they caught the gleam of steel flashing in their faces, gave way and fled. Two hundred and fifty dragoons, also posted on the right, seeing the artillery lost, became disheartened; they advanced to charge a large mass of Highlanders, but observing the disparity of numbers were seized with a panic, and galloped from the field.

This inauspicious commencement of the action damped the spirit of the infantry, and the panic spread from rank to rank; several companies, however, made resistance, and feats of valour were displayed by individuals and small parties; but all semblance of order was soon lost, and a complete rout ensued.

About four hundred of the royal forces were killed or wounded, and the prisoners, amounting to nearly twelve hundred men, were removed to Edinburgh, and afterwards to the Highlands.

The 55th had the following officers taken prisoners:— Lieut.-Colonel Sir Peter Halkett;* Captains Cochrane,

* The following testimony to the conduct of Lieut.-Colonel Sir Peter Halkett is contained in the "*Autobiography of the Rev. Dr. Alexander Carlisle, Minister of Inverkes,*" recently published. It may be necessary to remark that the above officer, who had succeeded to the baronetcy in this year, was Captain of a company as well as Lieut.-Colonel (see page 3), in order to explain the passage in which he is spoken of in reference only to the lower rank. "Sir Peter Halkett, a Captain in Lee's Regiment, "acted a distinguished part on this occasion; for, after the route, "he kept his company together, and, getting behind a ditch in "Tranent Meadow, he kept firing away on the rebels till they "were glad to let him surrender on terms." Sir Peter was after-

Chapman, and Tatton; Captain-Lieutenant Kennedy; 1745
Lieutenants Hewetson, Sandiland, and Dunbar; Ensigns Archer, Drummond, and Hardwick; Quarter-Master Wilson, and Dr. Young. This successful commencement of the rebellion caused numerous adherents to flock to the Prince's standard; several regiments were recalled from the Continent in October, and His Royal Highness the Duke of Cumberland proceeded to take the command of the Royal army. The young Pretender, elated with the capture of Carlisle, marched as far as Derby, whence, however, he commenced his retreat to the north on the 6th of December, as he found but few partisans in England to join in his expedition.

The battle of Culloden, fought on the 16th April, 1746, 1746
put an end to the rebellion, but the regiment was not present on that occasion.*

During the years 1747 and 1748 the 55th remained 1747
in Great Britain. The Treaty of Aix-la-Chapelle, which 1748
was signed on the 7th of October, 1748, terminated the "*War of the Austrian Succession.*"

In consequence of the conclusion of the war, several regiments were disbanded, and the 55th (as shown in page 2) henceforth became the 44th Regiment.

The regiment proceeded to Ireland in 1749, where it 1749
continued upwards of five years. 1750

King George II. conferred the colonelcy of the 44th 1751
Regiment on Colonel Sir Peter Halkett, Bart., on the 26th

---

wards dismissed on his parole, and was one of the five officers (the others being the Hon. Mr. Ross, Captain Lucy Scott, and Lieutenants Farquharson and Cumming) who refused, in February, 1746, to rejoin their regiments, notwithstanding the threat of the Duke of Cumberland, that they should be surperseded if they disobeyed his orders. The Government approved of their reply, that "His Royal Highness was master of their commissions, but not of their honour," and they accordingly remained in their regiments. Captain-Lieutenant Kennedy and Ensign Archer were retaken from the insurgents by the Angus Militia.

* Wolfe's Marines, at this period numbered the 44th Regiment, were at the battle of Culloden; but this corps was disbanded in 1878, and had no connection with the present 44th Regiment.

1751 of February, 1751, in succession to Colonel Lee, deceased. Sir Peter Halkett was Lieut.-Colonel of the regiment on its formation.

In the Royal Warrant of the 1st of July, 1751, for ensuring uniformity in the clothing, standards, and colours of the army, and regulating the number and rank of regiments, the facings of the 44th Regiment were directed to be *yellow*. The first, or King's colour, was the great Union; the second, or regimental colour, was of *yellow* silk, with the Union in the upper canton; in the centre of the colour was the number of the rank of the regiment, in gold Roman characters, within a wreath of roses and thistles on the same stalk.

1755 Major-General Braddock having been ordered to proceed to Virginia, the 44th and 48th Regiments were selected to form the force under that officer. They accordingly embarked from Ireland in January, 1755, and arrived in Virginia before the end of February.

The Peace of Aix-la-Chapelle was now about to be interrupted. The hostile spirit of the French first showed itself in aggressions on the British territory in North America. On Major-General Braddock's arrival in Virginia, with the 44th and 48th Regiments, it was determined to attack *Fort du Quesne*, on the Ohio, in the Province of Pennsylvania. This was part of a plan of assailing the French in America at all points, and, by one bold push, to repel them from all their encroachments. In order to preserve Oswego, and reduce Niagara, Shirley's and Pepperel's Regiments, which were raised for service in North America (afterwards disbanded), were ordered to march to Lake Ontario, and General, afterwards Sir William Johnson, Bart., proceeded in command of the provincial troops, with the intention of investing Crown Point, on Lake Champlain.

Many unforeseen and inconceivable difficulties attended this expedition against *Fort du Quesne*. Major-General Braddock was obliged to march his army across the Alleghany mountains, over a rugged, pathless, and unknown country, through unfrequented woods and dangerous defiles, which were rendered more perilous by the absence of any assistance whatever from the provinces; this appears in his correspondence with the British Ministry, complaining of the neglect and disaffection of all employed to supply necessaries for the troops, and setting forth the continual labour and fatigue of the soldiers, who had to cut out roads across steep mountains and rocks of an excessive height, and divided by rivers and torrents. 1755

The troops arrived safely at Fort Cumberland; and, being informed that the French, at *Fort du Quesne*, which they had recently erected on the river, near the confluence of the Monongahela, were expecting a reinforcement of five hundred regular troops, he resolved to push forward by forced marches. The General proceeded from Fort Cumberland on the 10th June, with fourteen hundred men and the greater portion of the ammunition and artillery, and arrived on the 8th of July within ten miles of the fort, without encountering any opposition. Colonel Dunbar, with eight hundred men, was left to escort the provisions, stores and baggage, as expeditiously as the nature of the service would permit.

While thus encamped, forty miles in advance of his corps of reserve, the General was exposed to all the stratagems and enterprise of an insiduous enemy. This was expressed to him in the strongest terms, especially by Colonel Sir Peter Halkett, of the 44th, who earnestly requested him to proceed with caution on such hostile and dangerous ground, and to order the Indians to reconnoitre, and act as scouts or advanced guards, in case of an ambuscade, for which the country was so well adapted. Unfortunately, no attention was paid to this excellent advice. The men were commanded to resume their march next day, without any steps to ascertain the situation of the enemy, or precautions to prevent surprise.

Having proceeded with this blind infatuation through a defile, the enemy being concealed behind the trees and bushes, the little army, of which the 44th formed part, about noon on the 9th of July, was surprised by a general

1755 fire on its front and flank. The advanced guard was formed by the Grenadiers of the 44th, under Brevet Lieut.-Colonel the Honourable Thomas Gage; twenty-two men only of their number were left standing after the first discharge. The enemy could alone be discovered by the smoke of their muskets, and great confusion was created by the tumultuous pressing on and crowding of the men behind, who earnestly requested instructions how to proceed. In this emergency several of them fell by the irregular fire of their comrades, as was ascertained afterwards, the bullets being of a larger size than those used by the foe. By this time the hundred Grenadiers were all killed save eleven, three of whom were mortally wounded. A party of these gallant fellows, placing themselves behind a large tree which had fallen down, used it as a breastwork, and made considerable havoc amongst the enemy. Most of the officers and men, who stood firm to the last, were sacrificed to the General's further mismanagement. Instead of directing a retreat, when his men were thus surprised, and taking steps to scour the avenues, lined by the enemy, with grape-shot from ten pieces of cannon he had with him, or ordering the Indians to advance in flanking parties against the hidden enemy, General Braddock remained upon the spot, his obstinacy increasing with the danger, until nearly all his officers and men about him were killed or wounded. After two hours and twenty minutes continued slaughter, rather than fight, the men, having fired away not only their own but likewise all the stock of cartridges they could find about the dead and wounded, this unequal contest was relinquished. The general, having had five horses shot under him, and a musket ball through his right arm and lungs, was carried off the field by the bravery of Lieut.-Colonel the Honourable Thomas Gage, of the 44th, and another officer, but survived only four days, his death being caused more by vexation and grief than by his wounds, which his surgeon declared were not mortal.

The loss of the British amounted to seven hundred men. The retreat was ably conducted by Colonel Washington, who had two horses shot under him in this action.

The officers suffered severely, the Indians being excellent

marksmen. Colonel Sir Peter Halkett, Bart.,* at the head of the 44th, fell at the first fire. Thus died, in the service of his king and country, this good and gallant officer, and amiable and accomplished gentleman; both in life and death a lasting honour to his family. 1755

* Lieutenant James Halkett, the youngest son, had rushed to the aid of his father, but, receiving at this instant his own death wound, fell lifeless, according to some accounts, across Sir Peter's body. The finding of their remains, covered with the leaves of three seasons, on the occupation of Fort du Quesne by the British, in November, 1758, is thus narrated by Allen Cunningham, in his life of Benjamin West, "*Lives of Painters*," vol. ii.:—

"West took up a musket—inspired with his enthusiasm, young "Wayne, afterwards a distinguished officer—and joining the troops "of General Forbes, proceeded in search of the relics of that gallant "army lost in the desert by the unfortunate General Braddock.

"To West and his companions were added a select body of "Indians; these again were accompanied by several of the Old "Highland Watch—the well-known 42nd—commanded by the most "anxious person of the whole detachment, Major Sir Peter Halkett, "who had lost his father and brother in that unhappy expedition. "Though many months had elapsed since the battle, and though "time, the fowls of the air, the beasts of the field, and wild men "more savage than they, had done their worst, Halkett was not "without hopes of finding the remains of his father and his brother, "as an Indian warrior assured him that he had seen an elderly officer "drop dead beneath a large and remarkable tree, and a young "subaltern, who hastened to his aid, fall mortally wounded across "the body. After a long march through the woods, they approached "the fatal valley. They were affected at seeing the bones of men, "who, escaping wounded from invisible enemies, had sunk down "and expired, as they leaned against the trees, and they were "shocked to see in other places the relics of their countrymen "mingled with the ashes of savage bivouacs.

"When they reached the principal scene of destruction, the "Indian guide looked anxiously round, darted into the wood, and in "a few seconds raised a shrill cry. Halkett and West hastened to "the place—the Indian pointed out the tree—a circle of soldiers was "drawn round it, whilst others removed the leaves of the forest "which had fallen since the fight. They found two skeletons—one "lying across the other—Halkett looked at the skulls, said faintly, "'It is my father!' and dropped senseless in the arms of his companions, "On recovering, he said, 'I know who it is by that artifical tooth.' "They dug a grave in the desert, covered the bones with a Highland "plaid, and interred them reverently. This scene at once picturesque

1755 Sir Peter's servant went up to his master to see if the wound was mortal, or in case he was dead to have the body decently disposed of; but the poor fellow at that instant was shot dead, and fell on his master's corpse, a worthy example of duty and fidelity.*

The other officers of the regiment, killed, were—Captains Charles Tatton and Henry Bromley, Captain-Lieutenant Richard Gethin; Lieutenants James Halkett, James Allen, and Robert Townsend.

The following officers of the regiment were wounded:—Lieut.-Colonel the Honourable Thomas Gage, afterwards so well known as the Commander of the Forces in Boston at

---

"and pious, made a lasting impression on the artist's mind. After "he had painted the 'Death of Wolfe,' he proposed the finding of "the bones of the Halketts, as an historical subject; and, describing "to Lord Grosvenor the gloomy wood, the wild Indians, the passionate "grief of the son, and the sympathy of his companions, said, he "conceived it would form a picture full of dignity and sentiment. "His Lordship thought otherwise. The subject which genius chooses "for itself, is, however in most cases, the best. The sober imagination "of West had here a twofold excitement—he had witnessed the "scene, and it was American—and, had Lord Grosvenor encouraged "him to embody his conception, the result would, I doubt not, have "been a worthy companion to the 'Death of Wolfe.'"

It must be added, that there was no Major Sir Peter Halkett in the 42nd Royal Highland Regiment, or in the Army in 1758. There was, however, a Captain Francis Halkett in the 44th at that time, and although the regiment did not form part of the force detailed against Fort du Quesne in the above year, yet that officer served with the expedition under Brigadier Forbes, and had the local rank of major. In "*Burke's Peerage and Baronetage*," Francis Halkett is shown as major in the Black-Watch, now the 42nd Regiment This is incorrect, but coupled with the above circumstances, would seem to point out that the remains were discovered by the second son, Major Francis Halkett, who did not long survive, as he died in 1760, then belonging to the 44th Regiment.

* These and other particulars relating to this sad affair have been gathered from a manuscript statement drawn up at the time by an American gentleman, Mr. Hamilton, and kindly lent by the present baronet, Sir P. Arthur Halkett, to the compiler of this record. As this account is apparently derived from authentic sources, and is so distinct, it is scarcely possible to believe but that it was the servant, and not the son, who died across Sir Peter's body.

the commencement of the American War; Lieutenants 1755
William Littler, William Dunbar, John Treby, Andrew Simpson, Robert Lock, Daniel Disney *(Adjutant)*, Quinton Kennedy, and Ensign George Pennington. Captain Francis Halkett, who escaped unhurt, proved himself, by his gallant bearing, worthy of his lamented father.

It was in this action that George Washington* first gave promise of that military talent which was in after years displayed *against* the British Army, in the American War, in which the 44th Regiment was employed. Gates, a name also celebrated in the American struggle, was likewise here.

On the 17th of July, the sick and wounded arrived at Fort Cumberland, and were shortly afterwards joined by Colonel Dunbar, with the remainder of the army.

Colonel Robert Ellison was appointed Colonel of the 44th Regiment on the 13th November, 1755, in succession to Colonel Sir Peter Halkett, Bart., killed in action.

On the 13th March, 1756, Major-General James 1756
Abercromby was appointed Colonel of the 44th Regiment, in succession to Colonel Robert Ellison, deceased.

Early in this year Louis XV. prepared a powerful armament for the capture of the Island of Minorca, and, on the 18th of May, Great Britain declared war against France. This was the contest known as "The *Seven Years' War.*"

In August, 1756, Oswego, on Lake Ontario, was taken

---

* Colonel Washington was Volunteer Aide-de-Camp to General Braddock: being seized with fever, his illness at length forced him to remain behind, and to relinquish the covered waggon in which he had hitherto been conveyed. The young officer rejoined the advanced division on the 8th July, and, although much debilitated, mounted his horse on the following day. Washington, in after years, often spoke of the display of the British troops on the eventful morning of the 9th, as being the most beautiful spectacle he had ever beheld.

Thackeray, in his popular work "The Virginians," has introduced, as a portion of the story, this disastrous action.

Colonel Washington commanded a brigade in the second expedition against Fort du Quesne, in 1758.

1756 by the French, and the garrison made prisoners of war. The 44th had proceeded on the 12th of that month to its relief, but some delay occurred owing to the interference of the provincial governors; while on the march, intelligence was received of its fall. After this success the French demolished the works at Oswego, and rejoined their army at *Ticonderoga*.

1758 The expedition against Louisburg, then the capital of the Island of Cape Breton, which had been deferred in 1757, proceeded thither in May, 1758, while the 44th Regiment was ordered to join a body of troops under Major-General James Abercromby, destined for the attack of *Fort Ticonderoga*. This force, which comprised the 27th, 42nd, 44th, 46th, and 55th Regiments, embarked on Lake George on the 5th of July, and landed on the following day near the extremity of the lake, whence the troops marched, through a wild and thickly wooded country, in four columns, upon *Ticonderoga*; the guides unfortunately mistook the route through the trackless woods, and, on the 6th of July, a skirmish ensued with a body of French troops, in which Brigadier-General George Augustus Viscount Howe (of the 55th) was killed. With this exception the British sustained but small loss, while the enemy had three hundred killed and one hundred and forty-eight taken prisoners. On the 7th of July, the troops being greatly fatigued by having been one whole night on the water, the following day constantly on foot, and the next night under arms, added to their being in want of provisions,—having dropped what they had brought with them in order to lighten themselves,—it was thought advisable to return to the landing place. About eleven in the forenoon Lieut.-Colonel Bradstreet, Deputy Quartermaster-General in America, with the 44th, six companies of the first battalion of Royal Americans (now the 60th Regiment), the bateau men, and a body of rangers and provincials, was sent to take possession of the Saw Mill, within two miles of *Ticonderoga*, which was soon effected. The enemy, before retiring, broke down the bridge and destroyed the mill. Lieut.-Colonel Bradstreet laid another bridge across, and the remainder of the troops took up their quarters there that night. No time was lost in making the attack. On the 8th of July the British appeared before the fort, which could only be approached on one side, which was strongly fortified, the other three sides being surrounded by water. Felled trees, with their branches outward, were ranged before the works, which were defended by between four and five thousand men. 1758

The engineer having reported that the intrenchment might be forced by musketry alone, Major-General Abercromby, unfortunately, determined to attack the place without waiting for the artillery, which, on account of the unfavourable nature of the ground, could not be easily brought up. A rumour also that the French were about to be reinforced with three thousand men confirmed the General in his resolution. After several repeated attacks, which lasted upwards of four hours, under the most disadvantageous circumstances, and with the loss of five hundred and fifty-one killed and one thousand three hundred and fifty-six wounded, orders were given to withdraw, and the troops retired to the camp occupied the night before. Early the next morning they proceeded to the camp at Lake George, and arrived there in the evening of the 9th of July. All the wounded officers and men that could be moved were then sent to Fort Edward and Albany.

Ensign William Fraser, of the 44th, was killed. No separate return of casualties has been preserved.

In the year 1759 it was proposed to assail the French in all their strong posts in Canada at once, so as to fall as nearly as possible at the same time upon Crown Point, Niagara, and the forts to the south of Lake Erie. A great naval armament, and a considerable body of land forces, under Major-General James Wolfe, were also simultaneously to attack Quebec by the River St. Lawrence. 1759

Lieut.-General (afterwards Lord) Amherst, who commanded the British forces in America, was to attack Ticonderoga and Crown Point by Lake George. The reduction of these forts would command Lake Champlain, where, having established a sufficient naval force, he was, by the

1759 River Sorel, which formed the communication between this Lake and the River St. Lawrence, to proceed to Quebec, and effect a junction with Major-General Wolfe.

The third of the grand operations was against *Fort Niagara*, near the celebrated falls of that name—a place of great consequence. The reduction of this fort was committed to Brigadier-General John Prideaux (55th), under whom Sir William Johnson commanded the provincials of New York and several Indians of the Five Nations, whose services, owing to the influence that gentleman had acquired among their tribes, had been secured on the British side. It was to this portion of the army that the 44th Regiment was attached.

The troops which had been appointed to proceed to *Niagara* arrived at the fort in July. The siege of the place had not been long commenced before Brigadier-General Prideaux was killed in the trenches, by the bursting of a cohorn. This occurred on the 20th of July, and the accident threatened to endanger the success of the operations; but Sir William Johnson, upon whom the command devolved, omitted nothing to continue the vigorous measures of his predecessor, and added everything his own genius could suggest.

Alarmed for the safety of the fort, the French collected all the troops they could draw from their posts about the lakes, and to these were joined a large body of Indians. The whole advanced to raise the siege, amounting in all to seventeen hundred men.

Sir William Johnson, on the 23rd of July, received intelligence of the approach of the enemy to relieve the fort, and immediately made a disposition to frustrate their designs. The guard of the trenches was commanded by Major John Beckwith, of the 44th, and, lest the garrison should sally out, and either attempt to surprise or overpower that guard, by which the British would have been hemmed in between two fires, the 44th Regiment, under Lieut.-Colonel William Farquhar, was posted in such a manner as to be able to sustain Major Beckwith.

The road on the left of the line, which led from the 1759 cataract to the fort, was occupied by the light infantry and picquets of the army on the evening of the 23rd of July. Early next morning these were reinforced by the Grenadiers and part of the 46th Regiment, the whole commanded by Lieut.-Colonel Eyre Massey, of the 46th, to whose judicious distribution of the troops, and the steadiness with which he received the enemy in front, while the Indians in British pay fiercely attacked them on the flanks, the success obtained was in a great degree attributable. The French were completely defeated, and all their officers made prisoners, among whom were Messieurs D'Aubry, De Lignery, Marin, and Repentini.

This action sealed the fate of *Fort Niagara*, which surrendered on the following day (25th of July), and Sir William Johnson, Bart., in his despatch to Lieut.-General Amherst, of that date, thus alluded to the conduct of the troops :—

" Permit me to assure you, in the whole progress of the " siege, which was severe and painful, the officers and men " behaved with the utmost cheerfulness and bravery."

Meanwhile the siege of *Ticonderoga* was prosecuted with vigour by the troops under Lieut.-General Amherst, and on the 25th of July the garrison blew up the fort and sailed to Crown Point, another fort on Lake Champlain, which place the French also abandoned, and retired down the lake to Isle-aux-Noix. *Crown Point* was occupied by the British on the 4th of August following.

The battle fought on the 13th of September, 1759, on the *Heights of Abraham*, in which Major-General James Wolfe was killed, led to the surrender of Quebec, which capitulated five days afterwards. This great success of the British troops ended the year in a most triumphant manner, and made the name of their Commander for ever memorable.

While the above operations were being performed, Lieut.-General Amherst found that the command of Lake Champlain was still an object of some difficulty, although the retreat of the French from Crown Point and Ticonderoga

1759 had left him master of Lake George. In October the troops embarked in boats, and proceeded a considerable distance along the lake; but the season became too far advanced for operations, which were postponed to the following year, and the force returned to Crown Point and Ticonderoga for winter quarters.

1760 The French endeavoured to regain possession of Quebec, and after the battle of *Sillery*, fought before that place on the 28th of April 1760, in which, from their superiority in numbers, they had the advantage, trenches were opened by them before the town. The 44th were not engaged in this action, and it is only adverted to here in order to preserve the thread of the narrative of the conquest of Canada, in which the regiment participated. The arrival of the English fleet in May allayed all fears for the safety of Quebec, and nothing now remained to cloud the prospect of the reduction of Canada by the united efforts of three British divisions, which, by different routes, were marching to attack those parts of the country that remained in the power of France.

A large column had been collected at Oswego by Lieut.-General Amherst, of which the 44th formed part. The grenadiers of the force, amounting to about six hundred men, were embodied, and the light infantry companies, consisting of the same number, were similarly incorporated. Thirty bateaux were provided for the 44th Regiment, and the troops embarked on the 10th and 11th of August, the rear-guard of the army being under the command of Brigadier-General the Hon. Thomas Gage.

Dispositions were afterwards made for the attack of *Fort Levi*, on *L'Isle Royale*, and, after two days' sharp firing, it surrendered on the 25th of August, and was immediately taken possession of by Lieut.-Colonel Massey, of the 46th, with three companies of Grenadiers. Some time was occupied in repairing this post and in fitting up vessels for conveying the troops down the St. Lawrence. Notwithstanding all precautions, nearly ninety men were drowned in passing the dangerous falls, and a great number of vessels broken to pieces. After a tedious voyage the British arrived, on the 6th of September, in sight of the Island of Montreal.

No time was lost in preparing the attack; the troops 1760
were immediately landed, and so admirably was the plan concerted, that Brigadier-General the Hon. James Murray landed from Quebec on that very day, and Colonel Haviland, with his force from Isle-aux-Noix, on the day following.

The French Governor-General, the Marquis de Vaudreuil, saw himself entirely enclosed, and was compelled to surrender the garrison of Montreal on the 8th of September. Thus was completed the conquest of Canada, which vast country has since continued under the dominion of Great Britain.

During 1761 and 1762 the 44th Regiment remained 1762
stationed in Canada, but was not engaged in any event of importance beyond guarding the acquisitions already obtained. In the former year the regiment was distributed at Trois Rivières, Chambly, St. Francis, and other places, but in October, 1762, the ten companies were concentrated at Montreal.

Negotiations for peace were commenced towards the close of the year 1762, and the preliminary articles were signed at Fontainebleau, by the Duke of Bedford, on the 3rd of November.

The Treaty of Fontainebleau was concluded at Paris on 1763
the 10th of February, 1763, the ratifications were exchanged on the 10th of March, and peace was proclaimed in London on the 22nd of that month. In this year the regiment, which had been reduced to nine companies, had four at Crown Point, four at Fort William Augustus, and one at Ticonderoga.

Towards the end of the year 1765 the 44th Regiment, 1766
which had been removed to Quebec, was relieved, and to
returned home. 1775

The regiment on its arrival was stationed in Ireland, in which country it continued until May, 1775, when the following circumstances occasioned its being embarked for the scene of its former services.

Serious disputes had arisen on the subject of taxation between the colonists in North America and the British

1766 Government. The passing of the Stamp Act, in 1764, was
to the first cause of irritation, but the spirit of discontent was
1775 partially allayed by its repeal in 1766. This feeling was again aroused in the following year by the Bill for levying duties on certain articles imported from England, which was repealed in 1770, with the exception of the duty on tea, that being retained as an assertion of the right of taxation inherent in the British Legislature. The cargoes of tea sent to Boston in 1773 having been emptied into the sea, an Act of Parliament was passed in 1774 for closing that port. Retaliatory measures were adopted by the colonists, and preparations were subsequently made for an appeal to arms

The first hostile collision between His Majesty's troops and the colonists in the unhappy contest which was soon to assume a most formidable character took place at *Lexington*, on the 19th of April, 1774.

The 44th Regiment embarked from Cork for North America on the 12th of May, 1775, to reinforce the troops at Boston, under General the Hon. Thomas Gage, who had formerly commanded the regiment.

During the night of the 16th of June the Americans commenced fortifying the heights on the Peninsula of Charlestown, named *Bunker's Hill*, and on the following day they were attacked by the British, and driven from their works.

The 44th Regiment shortly afterwards arrived in America, and was quartered at Boston.

Major-General the Honourable Sir William Howe, K.B., with the local rank of General in America, assumed the command of the British troops in that country, on General Gage returning to England in October.

1776 In March, 1776, Boston was vacated by the British, and the regiment proceeded thence to Nova Scotia.

From Nova Scotia the 44th sailed with the expedition to *Staten Island*, near New York, where the army was reinforced by several corps from Great Britain. The 23rd,

44th, 47th, and 64th Regiments formed the sixth brigade, commanded by Major-General Robertson. 1776

On the 4th of July, 1776, the American Congress issued their Declaration of Independence, abjuring their allegiance to the British Crown, and all hope of reconciliation became futile.

A landing was effected by the British on *Long Island* on the 22nd of August, and on the evening of the 26th the army was put in motion to pass a range of woody heights, which intersect the island, and to attack the Americans in position beyond the hills. The column under Major-General James Grant, of which the 44th formed part, was directed to advance along the coast with ten pieces of cannon, to draw the enemy's attention to that quarter. Moving forward at the appointed hour, this column fell in with the advanced parties of the Americans about midnight, and at daybreak on the following morning encountered a large body of troops formed in an advantageous position, defended by artillery. Skirmishing and cannonading ensued, and continued until the Americans, discovering by the firing at *Brooklyn* that the left of their army had been turned and forced, retreated in great confusion through a morass. The American Army, being thus driven from its positions with severe loss, made a precipitate retreat to the fortified lines at *Brooklyn.*

The 44th had Captain Andrew Browne and ten rank and file killed, Captain Primrose Kennedy, Lieutenant Thomas Browne, one sergeant, and seventeen rank and file wounded.

During the night of the 28th of August the Americans abandoned their fortified lines at *Brooklyn*, and retired across the East River in boats to New York. The reduction of *Long Island* was thus accomplished in a very few days, with little loss.

The regiment shared in the operations by which the capture of *New York* was accomplished; also in the movements by which the Americans were driven from *White Plains*, and in the reduction of *Fort Washington.*

1776 After the reduction of *Fort Washington* and of *Fort Lee*, on the opposite side of the North or Hudson River, the regiment continued the pursuit of the enemy across Jerseys, by Elizabeth Town, Raway, &c., towards Philadelphia. The 44th Regiment was afterwards stationed at Haarlem, a short distance from New York, and at the end of the year had one company at New York, seven at Hell-Gate, and two in New Brunswick.

1777 During the winter General Washington suddenly passed the Delaware River, and succeeded in surprising and making prisoners a corps of Hessians at Trenton, but he afterwards made a hasty retreat. Being reinforced, he again crossed the river, and took up a position at Trenton.

Extensive depôts were prepared by the Americans at *Danbury* and other places on the borders of Connecticut, and the 44th formed part of a body of troops which embarked from New York, under Major-General William Tryon, for the destruction of these magazines. The British arrived off Norwalk in the evening of the 25th of April, landed without opposition, and commenced their march for *Danbury*, whence the Americans fled as the British approached in the afternoon of the following day. As no carriages could be procured to bring off any part of the immense collection of stores at this place, the magazines were set on fire, and, owing to the spreading of the flames, the town was unavoidably burnt. This service accomplished, the British commenced their march back to the coast on the morning of the 27th of April, during which a body of Americans hung upon their rear, and at every eminence a corps of militia was found ready to oppose their march; but they attacked and routed their opponents, and in one of the skirmishes the American General Wooster was killed.

Arriving at *Ridgefield*, the British were opposed by a strong body of the enemy, under General Arnold, protected by intrenchments, which they had been for some time preparing. A few rounds, however, from the British artillery, and a gallant charge with the bayonet, routed the American force, and the King's troops halted at *Ridgefield* for the night.

Resuming the march on the following morning, the British were harassed by the enemy in their retrograde movement, and numerous skirmishes took place. Arriving at the *Hill of Compo*, contiguous to the place of embarkation, the Americans appeared in force, and commenced an attack with greater spirit and determination than before. The British troops confronted their numerous assailants, fired a volley, and charged with such impetuosity with the bayonet that their opponents were unable to withstand the shock, and retreated. The King's troops afterwards embarked without further molestation for New York. 1777

The 44th in this enterprise, on the 27th and 28th April, had three rank and file killed; Major Henry Hope, three sergeants, and twelve rank and file were wounded; one drummer and four rank and file were missing.

Afterwards, taking the field with the army in the *Jerseys*, the 44th Regiment was engaged in the operations designed to bring the enemy to a general engagement, but the Americans kept close in their fortified lines in the mountains. An expedition against the populous and wealthy city of *Philadelphia* was next undertaken. The regiment formed part of the third brigade, consisting of the 15th, 17th, 42nd, and 44th, under Major-General (afterwards Earl) Grey.

Embarking from Sandy Hook, the army sailed to the Chesapeake, and, proceeding up the Elk River, landed on the northern shore on the 25th of August. The American Army took up a position at *Brandywine*, to oppose the advance, and on the 11th of September the Royal forces moved forward to engage their opponents. The action proved decisive—the enemy was driven from his position, and forced to make a precipitate retreat. The 44th battalion companies were in reserve on this occasion, and were not engaged, but the flank companies took a conspicuous part in the action. Captain Benjamin Fish was wounded, but no separate account of the other casualties of the regiment has been preserved.

Information having been received that fifteen hundred

1777 Americans, commanded by General Wayne, were concealed in the woods three miles from the British quarters, to act against detached parties, Major-General Charles Grey proceeded during the night of the 20th of September with the first battalion of light infantry, the 42nd, and 44th Regiments, to surprise the enemy. The march was conducted with great secrecy, and, soon after midnight, the British approached the enemy's left, and at once bayonetted the picquets and out-guards. Guided by the light of the camp fires, they then rushed forward, and carried on the work of destruction with the bayonet. About three hundred Americans were killed and wounded, and eighty, including several officers, made prisoners. The remainder, favoured by the darkness of the night, escaped, leaving behind them their arms and eight wagons loaded with baggage and stores. The British, in this well-planned attack, had only one officer, one sergeant, and one private soldier killed, and a few men wounded. Gallantry in the troops, and ability in the commander, were fully manifested in this critical and arduous service.

The army continued its advance, *Philadelphia* was occupied, and the British took up a position at *Germantown*.

Making a forced march during the night of the 3rd of October, the American Army suddenly appeared before daylight on the following morning in front of *Germantown*, thinking to surprise the troops, and attacked the British outposts. The first assault was opposed by the second battalion of Light Infantry and the 40th Regiment, under Lieut.-Colonel Musgrave, posted at the head of the village. These corps were compelled to fall back, and the Lieut.-Colonel threw himself, with six companies of the 40th, into a large stone house, where he was attacked by an American brigade, aided by four pieces of cannon. This post was gallantly defended until Major-General Grey, at the head of the three battalions of the third brigade (of which the 44th formed part), his left being covered by the fourth brigade, under Brigadier-General James Agnew, Lieut.-Colonel of the 44th, by a vigorous attack repulsed the enemy with great slaughter. The 5th and 55th Regiments, advancing from the

right, engaged them at the same time on the other side of the 1777
village, and completed the defeat of the Americans in that quarter.

Brigadier-General James Agnew, Lieut.-Colonel of the regiment, was killed, and Major Henry Hope was promoted to the vacant Lieut.-Colonelcy. The other casualties of the regiment were—five rank and file killed, Ensign David Stack, one sergeant, and thirty-one rank and file wounded.

The regiment took part in the operations by which the British commander endeavoured to bring the enemy to a general engagement at *Whitemarsh,* and was afterwards quartered in the city of Philadelphia.

Having passed the winter in quarters in Philadelphia, 1778
the regiment, in the spring of 1778, furnished several detachments, which ranged the country in various directions to open communications for obtaining provisions. The regiment also shared in the fatigues and difficulties of the march of the army from that city through the Jerseys, in order to effect its return to New York, and its flank companies were engaged in repulsing the attack of the enemy on the rear of the column at *Monmouth Court-House,* near *Freehold,* in New Jersey, on the 28th of June, on which occasion Lieutenant Archibald Kennedy, of the grenadier company, was killed, and Lieutenant George Kelly, of the same company, was wounded.

An expedition, of which the 44th formed part, next proceeded against Bedford, on the Accushnet River, a noted place for American privateers. The troops were embarked under the command of Major-General Grey, and landed on the evening of the 5th of September. They overcame all opposition, and, having destroyed seventy privateers, with other shipping, together with an immense quantity of naval stores, &c., demolished the fort and blown up the magazines, returned on board the transports at noon on the following day. The troops then proceeded against Martha's Vineyard, destroyed the defences, captured three hundred and eighty-eight stand of arms from the militia, and compelled the inhabitants to deliver over three hundred oxen, ten thousand

1778 sheep, and a thousand pounds sterling, collected by the congress. After this success the regiment returned to New York.

The King of France having engaged to assist the Americans, the character of the war was changed, and the British Army was moved from Philadelphia to New York. Shortly after their arrival a powerful French armament appeared off that port. The enemy had a great superiority of numbers, but the enthusiasm in the British Navy and Army was unbounded, and the hour of contest was regarded with sanguine expectations. The enemy did not, however, venture to hazard an attack, but proceeded against Rhode Island. A numerous body of Americans co-operated in the enterprise, and invested Newport, the siege of which was subsequently raised by them.

At this period, in consequence of a powerful French armament menacing the British possessions in the West Indies, several regiments were embarked from North America for Barbadoes, but the 44th remained near New York, and towards the end of the year was thus distributed: one company in fort Knyphausen, seven at Laurel Hill, one at Southampton, and one at Jamaica, Long Island.

1779 In 1779 the Court of Spain commenced hostilities against Great Britain, and the Dutch followed the example.

On the 26th of March, 1779, Lieutenant Thomas Browne, Ensigns William Hamilton and Robert Robinson, of the 44th, embarked at Chatham, with drafts to complete the regiments in North America.

During this year the 44th Regiment was encamped in Staten Island; but in the autumn eight companies were removed to Paulushook, two remaining at the former station. The regiment was subsequently ordered to embark for Canada, but did not proceed to its destination for some months.

1780 On the 15th May, 1780, the regiment quitted New York for Canada, in which country it was stationed during the six following years.

Major General Charles Rainsford was appointed to the

Colonelcy of the 44th on the 4th May, 1781, in succession 1781
to General James Abercromby, deceased.

During the greater portion of this year the regiment was stationed at Quebec, but towards the end of 1781 it was distributed at St. Antoine, St. Charles, St. Dennis, and places near Montreal.

A letter, dated the 31st of August, 1782, conveyed to 1782
the regiment His Majesty's pleasure that county titles should be conferred on the Infantry, and the 44th received directions to assume the designation of the East Essex Regiment, in order that a connection between the corps and and that part of the county should be cultivated, it having been thought that such an arrangement would promote the success of the recruiting service.

On the 30th November, 1782, the preliminary Articles of Peace were signed at Paris, between Great Britain and the United States of America, and the treaty was concluded in the ensuing February.

In the following year (1783) peace was also concluded 1783
with France and Spain, and afterwards with Holland.

The regiment remained in Canada during the years 1784
1784 and 1785.

On the 13th of September, 1786, the 44th arrived in 1786
England from Canada, and the regiment occupied quarters in South Britain, but towards the end of October 1787, it was removed to Guernsey.

In April 1788, the regiment was relieved at Guernsey, 1788
and was stationed at Leeds until November, when it was moved to Tynemouth.

From this place the regiment proceeded to Scotland 1789
in May, 1789.

Four companies were detached from Scotland to the 1790
Isle of Man in July, 1790, and remained there during the following year.

The four companies at the Isle of Man proceeded to 1792
Ireland on the 17th of February, 1792, for which country the remainder of the regiment embarked at Portpatrick on the 22nd of that month.

During the year 1793 the regiment continued in Ireland. 1793

1793 Meanwhile Louis XVI. had been decapitated, and the progress of democracy menaced Europe with anarchy. On the 1st of February the National Convention of France declared war against Great Britain, whereupon the British Government prepared for hostilities, and a large army subsequently proceeded to the Continent, under the command of His Royal Highness the Duke of York, to join the Austrian and Prussian forces. The 44th, however, did not proceed to the seat of war until the year following.

In September, 1793, the flank companies of the 44th Regiment embarked for Barbadoes, in order to take part in the capture of the French West India Islands.

1794 A force was assembled at Barbadoes in the beginning of 1794, of which the flank companies of the 44th formed part, and early in February the expedition, under Admiral Sir John Jervis and General Sir Charles (afterwards Earl) Grey, sailed for *Martinique*, which island, after some sharp fighting, was taken possession of on the 22nd of March.

From Martinique the Grenadiers, under Prince Edward (afterwards Duke of Kent), the Light Infantry, under Major-General Dundas, and three other regiments, embarked on the 30th of March for *St. Lucia*, where they arrived on the 1st of April, and in three days the conquest of that island was effected. His Royal Highness Prince Edward, with his brigade of Grenadiers, and Major-General Dundas, with his brigade of Light Infantry, took possession of *St. Lucia* on the 4th of April. The flank companies of the 44th Regiment were afterwards employed in the reduction of the island of *Guadaloupe*. A determined resistance was made by the enemy: by the 20th of April however, the island was captured. Sir Charles Grey stated in his despatch that he "could not find words to convey an adequate idea, or to "express the high sense he entertained of the extraordinary "merit evinced by the officers and soldiers in this service."

At *Martinique* Lieutenant Henry Holland, of the 44th, was wounded, but, as the flank companies were formed into battalions of Grenadiers and Light Infantry, no separate return of casualties has been preserved.

The 44th having been ordered for service on the Continent, the battalion companies were removed from Ireland to England in April, 1794, and proceeded with the force, under General the Earl of Moira, on the 14th of June following. This force landed at Ostend on the 26th of that month, at which time the troops under the Duke of York, pressed by superior numbers, were retiring upon Antwerp. The Earl of Moira determined, therefore, not to limit the services of his reinforcement to the defence of Ostend, but to endeavour to join His Royal Highness, and after a tedious march, in the face of a superior and victorious enemy, whose troops were already overrunning the country in all directions, he arrived at Alost. While there, on the 6th of July, a body of French cavalry rode into the town, but was speedily chased out by the English Light Dragoons. 1794

After overcoming all difficulties, the Earl of Moira's corps joined the army under the Duke of York, at Malines, on the 9th of July. The Earl at this period returned to England, and the regiments which had composed his separate corps were incorporated with the army of the Netherlands, the 44th being brigaded with the 12th, 33rd, and 42nd Regiments, under Major-General Balfour. In August the regiment was in position near Breda, and in the beginning of September the forces under His Royal Highness retired to the vicinity of Bois-le-duc.

In the middle of September the enemy advanced in great force, and attacked the British posts on the right. The outpost at *Boxtel*, the advanced post of the allies, was forced, and the troops of Hesse D'Armstadt, which occupied it, sustained a severe loss. On the morning of the 15th of September the 44th, with the other regiments of their brigade, proceeded under Lieut.-Colonel the Honourable Arthur Wellesley (afterwards the celebrated Duke of Wellington), who then commanded the 33rd, to recover possession of *Boxtel.* The 44th had thus the honour of serving under England's greatest chief in his first brigade command, and in his first experience of active warfare. Lieut.-General (afterwards Sir Ralph) Abercromby, who commanded the force selected for this service, on finding the

1794 enemy so formidably posted, suspended the order for attack until further instructions were received from the Duke of York. His Royal Highness, adhering to his original determination, ordered an immediate movement of the troops, but at the same time all ulterior operations, after a first attack, were confided to Lieut.-General Abercromby's discretion. On clearing the village of Schyndel the mounted picquets of the enemy were perceived posted on a plain of considerable extent, skirted by a thick grove of fir trees. The British Dragoons advanced to drive them in, supported by two regiments of Foot Guards, with the 33rd and 44th, the 12th and 42nd being kept in reserve. The great force of their opponents prevented the attack being successful, but the British effected their retreat without molestation, except from a distant and ineffective cannonade. The loss of the regiment was limited to four men missing. Shortly afterwards the British troops retired beyond the River Maese.

The 44th was subsequently brigaded with the 8th, 37th, 57th, and 88th Regiments, under Major-General de Burgh, and was stationed near the Waal, to defend the passage of that river.

Meanwhile the flank companies of the regiment had remained at *Guadaloupe*, before which place a French armament appeared in June, 1794, to effect its recovery. The Grenadier and Light Infantry battalions in this island were composed of the flank companies of the 8th, 12th, 17th, 31st, 33rd, 34th, 38th, 40th, 44th, and 55th Regiments. On the 1st of September the Grenadier battalion mustered one hundred and fifty-two men fit for duty, having two hundred and eight sick, while the Light Infantry battalion mustered only thirty-three men fit for duty, having three hundred and eighty-two men sick.

After a most gallant defence, the post at *Berville Camp* was surrendered, on the 6th of October, to the French. The flank companies of the 44th did not form a portion of the garrison, but remained for some time longer in another part of *Guadaloupe*. The whole of the island, with the exception of *Fort Matilda*, having been regained by the French, this

fort was defended by the troops, under Lieut.-General 1794
Prescott, until the 10th of December, when it was evacuated by the British. The Lieut.-General stated in his despatch—"During the whole progress of this long and painful siege "the officers and men under my command conducted them-"selves in such a manner as to deserve my warmest praise, "bearing their hardships with the utmost patience and "fortitude, and performing their duty with the utmost "alacrity. The conduct of the whole garrison was such as "to entitle them to my best thanks, and I cannot particu-"larise the behaviour of any one officer without doing an "injury to the rest."

The waters of the Waal having become sufficiently 1795
frozen to bear an army with its *matériel*, the 44th and other regiments retired in January, 1795, through a country covered with ice and snow, the sufferings of the soldiers being of a most distressing character, which they endured with exemplary fortitude. The superior numbers of the enemy, the severity of the weather, and the defection of the Dutch, having rendered the evacuation of Holland indispensable, the British troops retired to Germany, and were quartered a short time in the duchy of Bremen. After halting a brief period in these comfortable quarters the 44th proceeded to Bremen-Lee, where the regiment embarked in transports, and in May arrived in England. The strength of the battalion companies was then eighteen officers, twenty-eight sergeants, twenty drummers, and four hundred and eighteen rank and file.

Only a brief interval from active service was permitted to the 44th, the regiment being selected to form part of an expedition which had been prepared for the deliverance of the French West India Islands from the power of Republicanism, and to reduce to obedience the insurgents of St. Vincent and Grenada. The troops, amounting to about twenty-five thousand men, in the highest state of efficiency, were embarked under the command of Lieut.-General Sir Ralph Abercromby, and convoyed by the Royal Navy, under Rear-Admiral Sir Hugh Christian. Unfortunately the voyage had been delayed until a very late period of the year. Three

1795 attempts were made to get under weigh, and each was prevented by the stormy weather, many ships being driven from their anchors and stranded.

Three hundred sail got under weigh on the 11th of November, when an accident to the Admiral's ship rendered the attempt of no avail. On the 15th another endeavour was frustrated by tempestuous weather. At length the fleet sailed, but it had scarcely got clear of the Isle of Wight when another severe storm obliged the vessels to return to Portsmouth. With great difficulty and exertion Admiral Christian collected the remainder of his convoy, and again sailed on the 9th of December; but four days afterwards again a storm destroyed many of the transports, and so scattered the fleet as to render a re-union impossible.

1796 The 44th having weathered the storm, which had dispersed the fleet, continued its voyage, and arrived at Barbadoes on the 28th February, 1796. Lieut.-General Sir Ralph Abercromby embarked in the "Arethusa" frigate, accompanied by such vessels as could be collected, and arrived at Barbadoes on the 14th of March following.

On the 22nd of April the fleet, with the troops destined for the attack of *St. Lucia,** of which the 44th formed part, sailed from Carlisle Bay, anchored in the evening of the 23rd in Marin Bay, Martinique, and on the evening of the 26th continued the voyage to *St. Lucia.* A landing was effected at several points on the 26th of April. Lieut.-Colonel Robert Liddell, of the 44th, on the 3rd of May obtained possession of the lower battery named Chapuis, and with the column under his command continued to hold it for some time, but, being unsupported, was placed in a very critical situation. The plan of attack having failed in execution, the troops retired to their original position, and the ships of war which were destined to enter the *Cul-de-Sac* returned to their anchorage. In this attack upon the enemy's batteries the 44th had four rank and file killed, and Captains George

---

* St. Lucia, which had been taken by the British in 1794, was recovered by the French in the following year.

Johnstone and William Tuffie, Lieutenant Edward Gregory 1796 (dangerously), and seventeen rank and file wounded ; sixteen rank and file were missing.

Every exertion was now made to carry out the original plan for the investment of *Morne Fortunée*, against which the batteries were opened on the 14th of May. On the night of the 17th the strong outwork of *La Vigie* was assaulted by the 31st Foot, and although the first battery was gallantly carried by storm, the regiment was eventually, after a severe struggle, obliged to withdraw.

On the 25th of May the island of St. Lucia capitulated, and the 31st, 44th, 48th, and 55th Regiments, under Brigadier-General (afterwards Sir John) Moore, were selected to occupy it. The loss of the 44th Regiment from the 3rd to the 24th of May was only one man killed.

The possession of the island was not destined to be a quiet one. Small bodies of French, who had deserted from the different fortresses at their capitulation, withdrew into the interior, and joined the runaway slaves and Caribs. Taking advantage of the impenetrable nature of the country, they formed themselves into bands, for the purpose of molesting the British and plundering the planters and other residents of the island. Brigadier-General Moore accordingly took the field, and penetrated with his force into the wildest quarters of the mountains, in order to eradicate these predatory bands. This service was of a most harassing nature.

During the remainder of the year 1796, and part of 1797 1797, the 44th remained in the West Indies, and the regiment, greatly reduced in numbers, embarked for England, and landed at Gravesend on the 31st of July. Whilst serving in the West Indies the following officers of the regiment died, namely :—Lieutenants John Charles Phipps, Samuel Tuffie, George Robert Stoney, and Adam Ogilvie ; Lieut.-Colonel Robert Riddell, Major Edward Wilson, Captain William Creagh, Captain-Lieutenant and Captain James Cooke ; Lieutenants William Waddell, John Wright, Donald M'Leod, Robert Chambers, and Jeffery Robinson ; Quarter-

1797 Masters Thomas Gardner and McCabe, and Surgeon Thomas Facon.

1798 The regiment did not remain long in England, as it embarked, under the command of Lieut.-Colonel David Ogilvie, at Portsmouth, on the 1st of October, 1798, in the transport "Weymouth," for Gibraltar. The numbers embarked were two field officers, seven captains, seventeen subalterns, five staff, forty-one sergeants, eighteen drummers, and three hundred and sixty-five rank and file.*

1799 During the years 1799 and 1800 the 44th remained in
1800 garrison at Gibraltar.

In October, 1800, the regiment quitted Gibraltar, having been selected to form part of an expedition to expel the French Army from Egpyt. It joined the force under General Sir Ralph Abercromby, at Malta, and was formed in brigade with the 18th, 30th, and 89th Regiments, under the orders of Brigadier-General John Doyle. The troops were embarked on the 20th of December for the Bay of Marmorice, in Asiatic Turkey, where the fleet arrived in nine days. Here the force was detained some time, owing to the tardy proceedings of the Turks, with whom a plan of co-operation had been arranged.

1801 On the 23rd of February, 1801, the fleet again put to sea, and on the 1st of March the expedition arrived off the celebrated city of *Alexandria*, and anchored in the bay of

---

* A small force was sent, under Major-General (afterwards Sir Eyre) Coote, in May, 1798, for the purpose of destroying the basin, gates, and sluices of the Bruges canal. The 44th did not embark on this service, but Major Donkin (afterwards General Sir Rufane Shaw Donkin, K.C.B.), of that regiment, highly distinguished himself in the action on the Sandhills, near Ostend, on the 20th of May, in which he was wounded, and, with the force engaged on that occasion, taken prisoner. This officer commanded the battalion of light companies, and in the official despatch the following testimony was borne to his conduct by Major-General Coote (severely wounded and taken prisoner) and Major-General Harry Burrard:—"We have, "Sir, however, the honour of mentioning to you Colonel Campbell, "of the 3rd Guards Light Infantry, and Major Donkin, of the 44th, "whose conduct, if anything could have protracted our fate, had "been equal to the difficulty of effecting it."

Aboukir. Notwithstanding all the exertions of the Navy 1801 under Admiral Lord Keith, the necessary arrangements for landing the troops could not be made until the 8th of March, chiefly in consequence of unfavourable weather. On the morning of that day a signal rocket was fired, and one hundred and fifty boats, laden with five thousand men, approached the shore. The calm silence of the morning, broken by the murmur of thousands of oars urging forward the *élite* of a gallant army, excited feelings of deep interest. The floating battalions drew near the shore, crowded with French troops, who, after sustaining severe loss, were forced to give way. No casualties were sustaine dby the 44th in this action.

In the evening after the action the victorious troops advanced three miles on the road towards *Alexandria*, and on the 12th the whole army moved forward, and came in sight of the enemy, then strongly posted on an advantageous ridge, his right on the canal of Alexandria, and his left towards the sea.

About six o'clock in the morning of the 13th of March the British advanced to attack the enemy's position in front of Mandora ; and after some sharp fighting the French were driven from their position, and forced to retreat over the plains to their lines on the heights before *Alexandria.* In this action the 44th formed part of the fourth brigade, and was brigaded with the 2nd, 30th, and 89th Regiments, the 18th having been posted to the second brigade.

The 44th had two rank and file killed ; Colonel Christopher Tilson,* Lieut. James Browne, and Ensign John Berwick,† two sergeants and twenty rank and file wounded. Lieutenant Brown afterwards died of his wounds.

---

* Afterwards General Christopher Chowne, who died on the 15th July, 1834. This officer was permitted to change his name from Tilson to Chowne on the 14th January, 1812.

† Ensign Berwick was wounded in the feet. A private in the regiment of a poetical turn named Burns, used to recite odes in honour of the victory before the officers at mess on each anniversary. In 1810 this couplet was received with great applause—

"The French that day darn't show their noses,
When gallant Berwick lost his toeses."

Captain Berwick was afterwards killed at Salamanca.

1801 After the ground had been duly reconnoitred, the British encamped in front of the enemy's lines, and on the morning of the 21st of March the French, issuing from their position, made an attack upon them ; but they encountered an opposition which they were unable to overcome, and the English Army was again victorious over the numerous veteran and vainly styled "invincible" troops of France.

In this protracted and well contested action, the 44th had one rank and file killed, Lieut.-Colonel David Ogilvie (mortally), one sergeant and fourteen rank and file wounded. The promotion did not go in the regiment, for Major and Brevet-Lieutenant-Colonel Kenneth McKenzie, of the 90th, was promoted, on the 26th of May following, to the vacant lieutenant-colonelcy.

No sooner had the enemy retired from this struggle, than the army became aware of the loss it had sustained in the Commander-in-Chief, General Sir Ralph Abercromby, who received a mortal wound at the commencement of the action, but which he concealed until the battle was decided. Sir Ralph Abercromby died on the 28th of March, and was buried at Malta.

Major-General (afterwards Lord) Hutchinson, who succeeded to the command, stated in his despatch, that it was impossible for him "to do justice to the zeal of the "officers and to the gallantry of the soldiers of this army," and "that in the arduous contest in which we are at present "engaged, His Majesty's troops in Egypt have faithfully "discharged their duty to their country, and nobly upheld the "fame of the British name and nation."

Shortly afterwards, a body of British and Turks traversed the country to the city of *Rosetta*, situated near the mouth of one of the great channels of the Nile, and distinguished for the beauty of its environs. This place was soon captured, but the fort of St. Julien held out. After its surrender, a strong division of the army advanced along the banks of this celebrated river to attack the French in Upper Egypt, but the enemy having been driven from *El Aft*, and from the fortified post of *Rahmanie*, retired upon *Cairo*.

Following the retreating foe up the country, the 44th arrived, in the early part of June, with the army in the vicinity of the celebrated pyramids of Egypt, and after a halt of several days advanced with the united British and Turkish forces for the seige of *Cairo*, the garrison of which, in a few days, surrendered on condition of being sent to France. 1801

The capture of the capital of Egypt added fresh laurels to the British arms, and the troops which had acquired these honours proceeded down the Nile to the vicinity of *Alexandria*, where, having driven in the French outposts, they vigorously commenced the siege of that place. In August, the 44th formed part of the first brigade of the line, under Major-General Ludlow, and was brigaded with the 25th, and the first and second battalions of the 27th Regiment. In the beginning of September, the garrison of *Alexandria*, surrendered, on condition of being forwarded to their own country.

In this manner was Egypt delivered from the power of France, and the British troops were rewarded with the thanks of Parliament, the approbation of their Sovereign, and the royal authority to bear on their colours the "SPHINX," with the word "EGYPT"; the Grand Seignior established the Order of Knighthood of the Crescent, of which the General Officers were made members; he also presented large gold medals to the field officers, and smaller ones, of the same pattern, to the captains and subalterns of the several regiments, which they were permitted by their Sovereign to receive and wear. As a further proof of the estimation in which the Sultan held these services, he ordered a palace to be built at Constantinople for the future residence of the British Ambassador.

The successful efforts of the British Fleet and Army were followed by a treaty of peace with France, by which Egypt was restored to the Ottoman Empire. The troops, as opportunities offered, were withdrawn from Egypt, and the 44th proceeded to Malta, where it arrived in October. It embarked for England in December, greatly reduced in numbers; indeed, there is a tradition of the corps, that the flank companies were represented by Sergeants R. Mackrell and J. Donaldson alone:—these non-commissioned officers

1801 were promoted to commissions in 1814, and subsequently died as lieutenants in the regiment.

1802 This treaty of peace was signed at Amiens on the 27th of March, 1802, but it was not destined to be of long duration, and only heralded the general European War which afterwards ensued.

On the 23rd of May, 1802, the 44th embarked at Bristol for Ireland, and arrived at Waterford on the 27th of May.

1803 In May, 1803, the war was renewed; Hanover was overrun by the French, and severed for a time from the British Crown. An immense flotilla was also assembled at Boulogne, for the invasion of England. The threat of invasion aroused the patriotism of the British people, and the most strenuous measures were taken to defeat the French ruler's designs. Numerous volunteer and yeomanry corps were formed in every part of the kingdom, and all party differences merged into one universal effort for the preservation of Great Britain. The "Army of Reserve Act" was passed in July, 1803, for raising men for home service by ballot, under which, and that of the "Additional Force Act," passed on the 14th of July of the following year, a second battalion was added to the 44th, and its history is introduced at page 54, after the services of the first battalion have been carried on to the end of the campaign in America in 1815.

On the 10th of October, 1803, the first battalion of the 44th embarked at New Geneva, Waterford, for England, and arrived at Portsmouth on the 17th of that month.

1804 The battalion remained in England during the year 1804.

1805 On the 26th of March, 1805, the first battalion embarked at Portsmouth for secret service, under the orders of General Sir James Henry Craig, K.B., and subsequently proceeded to Malta, where it arrived on the 19th of July following.*

* The flank companies continued in Sicily under Sir James Craig, until February, 1806, when they joined the head-quarters at Malta. Major Henry Powlett, of the 44th, however remained in Sicily, was present at the battle of Maida, on the 4th of July following, and was severely wounded. This officer, who had risen

During the years 1806 and 1807 the battalion remained 1806
in garrison at Malta. 1807

On the 28th May, 1808, the flank companies of the 1808
first battalion, consisting of twelve sergeants, six drummers, and one hundred and ninety-four rank and file, embarked from Malta for Sicily,* and on the 17th of September, the remainder of the battalion, amounting to forty-eight sergeants, sixteen drummers, and eight hundred and seventy-eight rank and file, likewise proceeded to the same destination.

Upon the decease of General Charles Rainsford, the 1809
colonelcy of the 44th was conferred by His Majesty on General Sir Thomas Trigge, K.B., from the 68th Regiment, the appointment being dated 27th of May, 1809.

In June, 1809, Lieut.-General Sir John Stuart, Commanding-in-chief in the Mediterranean, resolved to menace the capital and the kingdom of Naples, as a diversion in favour of the Austrians, who were contending against numerous difficulties in their war with France. The first battalion of the 44th, commanded by Lieut.-Colonel Arthur Brooke, formed part of this armament, and embarked from Sicily on the 11th of June. With the first battalion of the 27th, it formed the brigade under Brigadier-General John Oswald. After menacing a considerable extent of coast, the

---

to the rank of major in the 20th Foot, was appointed to the 44th on the formation of the second battalion, in July, 1804. While serving with his former regiment in Holland, during the campaign of 1799, he was twice wounded in the actions at Crabbendam and Egmont-op-Zee, on the 10th of September and 2nd of October of that year. His services with the light infantry battalion (consisting of the light companies 20, 27, 35, 58, 61, 81st, and Watteville's) at Maida, gained him the Lieut.-Colonelcy of the 7th Garrison Battalion; he was subsequently appointed Captain of Carisbrock Castle. Lieut.-Colonel Powlett died in February, 1817.

* The flank companies of the several regiments, in Sicily, were formed into battalions of grenadiers and light infantry, and remained thus embodied for some time. They were a fine body of men, and there existed great pride and jealousy among the several companies to keep up the credit of each corps. The songs and merry days passed with the two battalion messes were long afterwards remembered by the officers.

1809 island of *Ischia*,* celebrated for the beauty of its scenery, and situated in the Bay of Naples, about six miles from the coast, was attacked. A landing was effected in the face of a formidable line of batteries, from which the enemy was speedily driven. The siege of the castle was next undertaken, and in a few days the garrison was compelled to surrender. On being summoned, the island of *Procida* capitulated, and the battalion landed there on the 25th of June, from whence it was withdrawn in August following, and on returning to Sicily, was stationed at Messina.

Two valuable islands were thus rescued from the power of the Grand Duke of Berg, General Murat, upon whom the Emperor Napoleon had, in the preceding year, conferred the sovereignty of Naples, in succession to Joseph Bonaparte,

---

* Colonel Burney, who served as a subaltern at the capture of this island and that of Procida, affords the following description of the uniform of the 44th, on his joining it in 1808. The officers wore large cocked-hats, leather breeches, and long boots above the knee, like dragoons, with powder and long tails, the curl of which was generally formed of some favourite lady's hair, no matter what the colour might be. The evening dress was grey cloth tights, with Hessian boots and tassels in front. The facings of the coat were buttoned back, and every one was powdered, and correctly dressed before sitting down to dinner. For duty, officers and men wore white cloth breeches, black cloth leggings or gaiters, with about twenty-five flat silver buttons to each, and a gorget, showing the officer was on duty. At Malta, as in other garrisons, officers for duty were regularly examined, that their buttons and swords were quite bright, if not they were turned back, and the one in waiting brought forward. Members of courts-martial were sent back by the President if they had not their gorgets on, and their duty dress and hair properly powdered. To appear out of barracks, without being in strict regimentals and swords, was never dreamt of. The poor soldiers, ordered for duty, were excused the adjutant's drill, as they took some hours to make themselves up to pass muster for all the examinations for guard-mounting, with pomatum (sometimes a tallow candle), soap and flour, particularly the men of flank companies, whose hair was turned up behind stiff as a ramrod. The queues were doomed by General Order from the Horse Guards, dated 20th July, 1808. The officers wore flashes, made of black ribbon, instead of a tail, attached to the collar of the coat behind, to distinguish them as flankers. This custom has been for years preserved in the 23rd Royal Welsh Fusilers.

nominated by his brother to be King of Spain; one thousand 1809
five hundred regular troops, with one hundred pieces of ordnance, were captured during these operations. An attempt was next made to reduce the castle of *Scylla*, but the large force which the enemy possessed in Calabria, rendered this impracticable.

A detachment of the 44th, consisting of ten sergeants, three drummers, and one hundred and eighty-six rank and file, with a proportion of officers, embarked from Sicily on the 22nd of September, 1809, and proceeded to the Ionian Islands with the expedition under Brigadier-General John Oswald, Lieut.-Colonel of the 35th Regiment, for the purpose of delivering them from the power of France; but this detachment, having been relieved by one from De Rolle's regiment, returned to Sicily in March of the following year.

In the summer of 1810, General Murat assembled 1810
upwards of one hundred heavily-armed gun-boats, a number of lighter ones, with about four thousand transport boats, and brought thirty thousand troops to the coast of Calabria for the purpose of invading Sicily. A landing was effected during a dark night, between the 17th and 18th of September, about seven miles to the southward of Messina. This attack was repulsed, and about one thousand men were taken prisoners; after this futile attempt, Sicily remained unmolested. During these operations the 44th was stationed at Messina.

The battalion was removed from Messina to Salvadore 1811
dei Greci in March, 1811, and remained there during its continuance in Sicily.

On the 14th of August the battalion embarked for Malta, where it arrived on the 21st of the month.*

* While the first battalion was stationed at Malta, Lieutenant George Clarges Hill, of the grenadier company, a very fine young man, met at a ball the daughter of a Maltese merchant; he danced with her, and the next morning, at daylight, she came to his quarters, having quitted her home for the purpose of following him. The worthy subaltern lost not a moment in persuading her to return to her parents, and having at length succeeded, sent his servant to see her safely back. This honourable trait was long remembered in the regiment, and rendered the corps much respected in the island.

1812 The flank companies embarked from Malta, for Sicily on the 4th of January, 1812, but returned to the former island on the 10th of that month. Five companies proceeded to Sicily on the 4th of July, and returned to Malta on the 6th of February of the following year.

1813 On the 24th of March 1813, five companies proceeded from Malta to Sicily, and arrived there in the same month. These were followed, on the 10th of May, by the remaining companies, which arrived at Sicily on the 24th; they did not land, but sailed on the 9th of June, and reached Minorca on the 25th; proceeding from thence to the coast of Spain, they landed in the Bay of Saloo on the 3rd of August, and marched the same day to join the army before *Tarragona.*

The left wing continued at Palermo until the 16th of August, 1813, when it proceeded to the east of Spain, and joined the army on the 7th of September, at Villafranca.

Tarragona was occupied by the British, the French under Marshal Suchet having blown up the defences of that place on the night of the 18th of August, and retired towards Barcelona. Lieut. General Lord William Bentinck continued in command of this division of the army until the 23rd of September, when his Lordship embarked for Sicily, where fresh changes, injurious to British policy, required his presence; he was succeeded by Lieut.-General William Clinton.

1814 The first battalion proceeded from Tarragona to Villafranca in March, 1814, and was afterwards removed to Saragossa. In April hostilities were terminated by a treaty of peace. Napoleon abdicated the throne of France, the Island of Elba being ceded to him in full sovereignty, with the imperial title for life, and a pension payable by the Government of France. On the 3rd of May Louis XVIII. entered Paris, and ascended the throne of his ancestors.

Both battalions were afterwards permitted to bear the word "PENINSULA" on the colours, in commemoration of their services in Spain.

General John Earl of Suffolk, from the 70th Regiment,

was appointed Colonel of the 44th on the 12th January, 1814
1814, in succession to General Sir Thomas Trigge, K.B., deceased.

Although tranquility, after a lengthened struggle, had been restored to Europe, the first battalion of the 44th was only allowed a few days of repose when it received orders to embark for another scene of action, and was destined again to fight the battles of its country beyond the Atlantic.

During the Peninsular War the decrees of Napoleon to destroy the commerce of Great Britain were followed by regulations designed by this country to counteract the enemy's plans. These regulations, with the pressing of British seamen on board of American ships, occasioned a war between England and the United States. This was the service on which the battalion was about to be employed.

Upon the army on the east coast of Spain being broken up the 44th marched in April, 1814, to join the force in the south of France. Proceeding through Bayonne, on the 15th of May, the battalion arrived at Bordeaux, and embarked on the 29th at Pouillac, on the River Garonne, quitting the coast of France early in June.

The expedition, consisting of the 4th and 44th, about eight hundred bayonets each, the 85th, about six hundred bayonets, with a brigade of artillery and a detachment of sappers and miners, was under the command of Major-General Patrick Ross. The naval part of the expedition, consisting of one seventy-four, two sixty-fours, five frigates, and two bomb-vessels, was under the orders of Rear-Admiral Malcomb.

After twenty days' sail the fleet approached the Azores, the high land of St. Michael's appearing like a blue cloud rising out of the water. As the shipping drew near the troops were delighted at the view of the numerous villages, seats, and convents ornamenting the beach, and the groves of orange trees and green pasturage, with lofty mountains rising behind. In a short time the fleet again put to sea, and anchored on the evening of the 24th of July opposite the tanks at the Island of Bermuda, where the expedition was joined by the 21st Fusiliers, mustering nine hundred

1814 bayonets. Here the fleet remained, taking in stores and establishing a magazine for the future supply of the expedition until the 3rd of August, when it once more put to sea, and, directing its course towards North America, entered the Bay of Chesapeake, where it was joined by reinforcements, Rear-Admiral Cockburn assuming the naval command. On the arrival of this squadron a powerful American flotilla fled for refuge up the Patuxent River, and was followed by the British Fleet. In order to ensure the destruction of the enemy's vessels, the troops were directed to land. On the 19th of August the stream was suddenly covered with boats crowded with soldiers, and by three o'clock in the afternoon the army was in position about two miles above the village of St. Benedict, on the right bank of the Patuxent. The troops were divided into three brigades—the first, consisting of the 85th Regiment, with the light companies of the 4th, 21st, and 44th, a company of Marines, and a party of disciplined negroes, was commanded by Colonel Thornton ; the second, consisting of the 4th and 44th Regiments, was commanded by Colonel Arthur Brooke, of the latter corps ; and the third, consisting of the 21st Fusiliers and a battalion of Marines, was commanded by Colonel Patterson. For want of horses only one six-pounder and two small three-pounders were brought on shore.

The army remained in position near St. Benedict until four o'clock in the afternoon of the following day, when, the bugles sounding, the regiments turned out in marching order, and proceeded in the direction of Nottingham, a town on the banks of the Patuxent. This they found deserted, while the appearance of the furniture, and in some places the bread left in the ovens, showed it had been hastily abandoned. On the 22nd the army proceeded to the delightful village of Upper Marlborough, situated in a well-cultivated valley about two miles broad. During the march loud explosions were heard, which proved to be the enemy destroying his flotilla, to prevent its falling into the hands of the British.

Thus far the object of the expedition had been accomplished, but, as the army had advanced within sixteen miles of *Washington*, and the enemy's force was ascertained to be such as would authorise an attempt to carry that capital, the troops moved forward on the 23rd of August. They had scarcely proceeded three miles when the advance guard encountered a party of American riflemen, who maintained a sharp contest before they gave way; and, arriving at a point where two roads meet, the one leading to Washington and the other to Alexandria, twelve hundred Americans and some artillery appeared on the slope of an opposite height. The army turned along the road leading towards Alexandria, and the Americans fled before the detachment sent against them. Having deceived the enemy respecting the real design of the expedition, the route was changed, and the troops proceeded in the direction of Washington. 1814

About noon on the 24th of August a heavy cloud of dust was seen to arise at a distance, and the British troops, turning a sudden angle in the road, and passing a small cluster of trees, discovered above eight thousand American infantry, with a numerous artillery and three hundred dragoons, commanded by General Winder, occupying a formidable position beyond the village of *Bladensburg*, awaiting the advance of their opponents.

The British, though not half so numerous as their adversaries, advanced boldly to the attack. On entering the village their opponents' artillery opened a tremendous fire, and, as the light brigade passed the bridge across the eastern branch of the Potomac, whence a road ran through the enemy's position to the capital, numbers fell before a heavy fire of musketry and artillery. The survivors having gained the opposite side of the river, carried a fortified house which commanded the bridge, then dashed into the thickets on the right and left of the road, and drove back the American riflemen, who fled with such precipitation that they drew their first line into disorder, which latter fell back in confusion, leaving two guns on the road. The British Light Infantry, throwing off their knapsacks, pushed forward in extended order to attack the second line, but a heavy fire of musketry and artillery causing a check, and the Americans advancing in force to recover the lost ground, drove the first

1814 brigade back to the thickets on the brink of the river, where an obstinate fight was maintained. Meanwhile the second brigade passed the river, and the 44th, moving to the right, turned the enemy's left flank, while the 4th, emulating their gallant companions, advanced in firm array, preceded by a flight of rockets, to charge its right, which was broken and driven from the field. Many of the American sailors, who acted as gunners, were bayoneted, and eight guns were captured. The American infantry fled in dismay, and, diving into the recesses of the forest, were quickly beyond the reach of their pursuers, while their cavalry, turning their horses' heads, galloped off the field. Thus in one hour was the battle won, and the third brigade, which had formed the reserve, pushed forward with rapid strides for *Washington.*

The casualties of the 44th were confined to one sergeant and thirteen rank and file killed, and thirty-five rank and file wounded.

Major-General Robert Ross, in his official despatch, thus expressed his approbation of the conduct of the regiment :—

" I have to express my approbation of the spirited " conduct of Colonel Brooke and of his brigade. The 44th " Regiment, which he led, distinguished itself under the " command of Lieut.-Colonel Mullins. The gallantry of the " 4th Foot, under the command of Major Faunce, being " equally conspicuous."

Colonel Arthur Brooke, mentioned in the above extract, was Lieut.-Colonel of the 44th, and Brevet Lieut.-Colonel the Honourable Thomas Mullins was a captain in that regiment. The former was second in command of the army under Major-General Ross.

In consideration of the conduct of the 44th in this action, it was authorised to bear the word " BLADENSBURG " on the regimental colour and appointments.

The three British regiments which had thus defeated about nine thousand adversaries (three times their own number) halted a short time on the field of battle to re-form their ranks. After a brief halt the 44th, with the remainder of the first and second brigades, moved towards Washington,

where the third, having already arrived, had commenced the 1814
destruction of the arsenal, docks, magazines, and other public property. As it was not the intention of the British Government to attempt permanent conquests in this part of America, and as it was impossible for so small a force to establish itself in the enemy's capital, the destruction of public property, which, by the custom of war, is the just spoil of the conqueror, was completed, and the army marched back to St. Benedict, where it re-embarked without molestation.

After remaining a few days in the Patuxent River, the fleet weighed anchor; the coast was menaced at several points, and the shipping approached so near the shore at Annapolis that the inhabitants were discovered flying from their houses; wagons loaded with furniture were seen hurrying along the roads, alarm guns firing, beacons blazing, and the people apparently oppressed with all the horrors of doubt and apprehension.

Baltimore, having been selected as the point of attack, the fleet hastened, under a heavy press of sail, towards the river upon which it is built. During the night of the 11th of September, the troops cooked three days' provisions, and each man received eighty rounds of ammunition. At three o'clock on the following morning the boats were lowered; a landing was effected at North Point, thirteen miles from Baltimore, and the army moving forward, a division of Americans fled from an intrenched position they were preparing, across a neck of land towards which the British troops were advancing. About two miles beyond this post the country was closely wooded, and the enemy's riflemen opening a sharp fire from behind the trees, Major-General Ross rode forward to ascertain the disposition and number of the opposing force, and exposing himself among the skirmishers, was mortally wounded. Upon the fall of this gallant officer, the command devolved on Colonel Brooke, of the 44th. The army moving forward, found itself in a few moments in front of a strong position, near Nip Church, in Godly wood, occupied by six thousand Americans, with six pieces of artillery, and some cavalry. The light brigade immediately extended, and driving in the enemy's skirmish-

1814 ers, menaced the whole front of their position. The 44th, a party of seamen, and the marines of the fleet, formed line behind the light infantry; the 21st Fusiliers and the second battalion of Marines formed column in reserve, the soldiers resting themselves on the ground, while the 4th Foot (King's Own) moved to the right, along some hollow ways and woodlands, and gained, unperceived, a thicket on the enemy's left flank. Meanwhile, the deep tones of the artillery echoed in the woods, and the instant the 4th gained the thicket, the charge was sounded and repeated by every bugle of the force; the soldiers started from the ground, and moving forward with a firm and resolute tread, in the face of a shower of grape and canister-shot, approached their adversaries, who raising a loud shout, opened a heavy fire of musketry. This was answered with a British cheer, a volley, and a rush forward at the double with the bayonet; then the Americans, seeing the ranks of gleaming steel close upon them, faced about and fled into the thick woods in dismay, leaving two pieces of cannon behind them. Fifteen minutes had sufficed to decide the fortune of the day. Several hundreds of the enemy's killed and wounded lay scattered over the field; many fugitives were cut off and made prisoners, and a number of American riflemen, discovered in trees, which they had climbed, to enable them to take sure aim and escape danger, were shot on their perches by the British soldiers, who did not consider these tactics fair play.

Colonel Brooke stated, in his dispatch, that "the "Honourable Lieut.-Colonel Mullins deserved every appro-"bation for the excellent order in which he led that part of "the right brigade, under his immediate command, while "charging the enemy in line."

The loss sustained by the 44th on the 12th of September, when in action with the enemy near *Baltimore*, consisted of eleven rank and file killed; Brevet-Major John Cruice wounded slightly; Captain Hamilton Greenshields dangerously (died of wounds); Captain George Clarges Hill, Lieutenant Richard Cruice, and Ensign James White, severely; five sergeants, and seventy-eight rank and file were wounded.

Halting on the field of battle, the bivouac fires were lighted, and the victorious army reposed for a short time. Two hours after midnight the soldiers were again under arms; at dawn they resumed their march, and in the evening arrived at the foot of the range of hills in front of *Baltimore*, where the American army, consisting of upwards of fifteen thousand men, appeared, occupying a chain of palisaded redoubts, connected by breast-works, and defended by a numerous train of artillery. Trusting to the innate valour and excellent discipline of his little army, which did not amount to one-third of the number of the opposing host, Colonel Brooke made arrangements for storming the hills after dark; but having received intimation from the commander of the naval forces that the entrance of the harbour was closed up by vessels sunk for that purpose, and that a naval co-operation against the town and camp was impracticable, the enterprise was abandoned. The troops accordingly, on the 14th of September, withdrew to a distance of three miles, and then halted to see if the Americans would venture to descend the hills; but the latter, though so superior in numbers, had no disposition to quit their intrenchments, and the British forces, retiring leisurely to their ships, re-embarked. 1814

The armament remained a short time on the coast, and information having been received of the formation of an American camp, a few miles from the river Potomac, the 44th, with the remainder of the second and third brigades were landed on the night of the 4th of October, and pushed forward to attack the enemy, who, however, having received notice of the movement, had fled. The battalion, therefore, returned on the 5th, and the season having arrived when active operations could no longer be continued in the Chesapeake, the fleet sailed for the West Indies, and anchored, on the 31st of October, in Port Royal harbour, Jamaica, the troops remaining on board, while the vessels took in a supply of provisions.

While in the West Indies, reinforcements arrived, and Major-General John (afterwards Lord) Keane took command of the expedition.

1814 The next enterprise taken was of a most difficult character, namely, an attempt on *New Orleans*, a place of some note, on the eastern bank of the great river Mississippi, one hundred and ten miles from the Gulf of Mexico, and so situated that the approach of a hostile force was almost impossible. The fleet having put to sea, anchored on the 10th of December off the coast of Louisiana, opposite the Chandeleur Islands, where the troops were removed into light vessels, and entering Lake Borgne on the 13th, five of the enemy's large cutters, mounting eleven guns each, were captured by a flotilla of launches and ships' barges. Having proceeded a short distance along the lake, all the vessels ran aground; the soldiers were then conveyed 20 miles in open boats, during a heavy rain, to a barren spot called Pine Island, a swamp with a piece of firm ground only at one end. Here the regiments remained without tents or huts, exposed to heavy rains by day, and to frosts by night, until the 22nd of December, when the first division proceeded in open boats to a desert spot, about eight miles below New Orleans, where the regiments landed, and marched to a field on the banks of the Mississippi. The 21st, 44th, and 93rd regiments, under Colonel Brooke,* and a large proportion of artillery under Major Munro, were embarked in small vessels. The Americans, under Major-General Jackson, made a night attack on the British, but were repulsed. Major-General Keane stated, in his dispatch, that "Colonel Brooke was "entitled to every praise for securing our right flank." The 44th was not engaged on the 23rd and following day. Before dark on the 24th of December, the whole army was in position, and Major-General the Honorable Sir Edward Packenham here joined to take command, accompanied by Major-General Gibbs.

Notwithstanding the repulse given to the enemy, the troops had been unable to return to their camp, as it was completely commanded by the fire of an American schooner, and no provisions could be procured. The army afterwards

* Colonel Brooke, Lieut-Colonel of the 44th, was, on the extension of the Order of the Bath, nominated a Companion thereof on the 4th of June, 1815.

moved forward, but encountered many local difficulties, and 1814
the Americans having assembled a numerous force, in extensive fortified lines and batteries, with armed vessels on the river, the advance was checked, and some loss was sustained.

From the 25th of December, 1814, to the 5th of 1815
January, 1815, the 44th had Lieutenant John Blakeney and one private killed, five rank and file wounded, and one missing.

Arrangements were made for attacking the enemy's fortified lines at daybreak on the 8th of January. On the left bank of the Mississippi the position of the Americans was simply a straight line of front of about one thousand yards, with a parapet, the right resting on the river and the left on a wood which had been made impracticable for any body of troops to pass. This line was strengthened by flanking works, and had a canal of general depth of about four feet, but not altogether of an equal width; it was supposed to narrow towards the enemy's left, and eight heavy guns were in position on this line. The Mississippi at this point was about eight hundred yards across, and on the right bank was a heavy battery of twelve guns, which enfiladed the whole front of the position of the left bank. Preparations were made by the British, with very considerable labor, to clear out and widen a canal that communicated with a stream, by which the boats had passed up to the place of disembarkation, so as to open it into the Mississippi, by which means troops could be got over to the right bank, and the co-operation of armed boats secured.

The assailing of the enemy's lines in front was entrusted to the brigade composed of the 4th, 21st, and 44th, with three companies of the 95th (Rifles), under Major-General Gibbs; and to the third brigade, consisting of the 93rd, two companies of the 7th Royal Fusiliers, 43rd, and 95th Regiments, under Major-General Keane. The attacking columns were to be provided with fascines, scaling ladders and rafts, the whole to be at their stations before daylight.

This plan was, however, partly disconcerted by the tardy arrival of the boats; and at the moment of attack the scaling ladders and fascines had to be sent for. Daylight arriving,

1815 the troops were visible to the enemy, who opened a tremendous fire, with dreadful execution. Major-General the Honourable Sir Edward Pakenham, G.C.B., having galloped to the front to encourage the men, was shot on the top of the glacis. Major-Generals Gibbs and Keane were borne from the field dangerously wounded; the former died of his wounds the following day, but the latter recovered. Success being found impracticable, the troops were withdrawn from the unequal contest, and no prospect remaining of capturing *New Orleans*, they returned on board the fleet.

The 44th suffered severely. Lieutenant Rowland Davies and Ensign Matthew MacClosky, one sergeant, and thirty-two rank and file were killed. The wounded were Captain Henry Debbeig (Lieut.-Colonel) and Lieutenant William Maclean, slightly; Lieutenants Robert Smith, Henry Brush, Richard Phelan, and William Jones, severely; Ensigns James White (died of wounds), Robert L. Haydon, and John Donaldson, severely; five sergeants, and one hundred and forty-nine rank and file. Lieutenant William Knight, two sergeants, one drummer, and seventy-six rank and file were taken prisoners.

After the failure at New Orleans, it was agreed that operations should be carried towards Mobile, before which place a demonstration had been made in the previous September. Major-General Lambert, on whom the command had devolved, proceeded on the service with the 4th, 21st, and 44th Regiments and some engineers and artillery. A landing was effected on the 7th of February, three miles from Fort Bowyer, on a point of land commanding the entrance of the harbour. Brevet Lieut-Colonel Debbieg, of the 44th, commanded the advanced column, before which the enemy fell back on the Fort. Siege operations were commenced on the following day, and on the 11th, preparatory to opening fire, Major-General Lambert summoned the defenders to surrender. The American commander seeing that resistance was useless, agreed to his terms, and the garrison, three hundred and sixty six strong, marched out, laid down their arms, and were embarked on board the fleet. Further hostilities were terminated by a treaty of peace. 1815

The first battalion of the 44th disembarked at the Cove of Cork on the 24th and 29th of May, 1815, and was stationed in Ireland for nearly seven years.

Brevet Lieut.-Colonel the Honorable Thomas Mullins, Captain in the 44th, was tried by a general court-martial, held at the Royal Barracks Dublin, on the 11th of July, and continued, by adjournments, to the 1st of August following, "for having on the 8th of "January shamefully neglected and disobeyed the orders "received from the late Major-General Gibbs to collect "the fascines and ladders, and to be formed with them "at the head of the column of attack at the time directed, "and in disobedience of the said orders, suffering the "regiment under his command to pass the redoubt where "the fascines and ladders were lodged, and remaining at "the head of the column for half an hour and upwards, "without taking any steps to put the 44th Regiment in "possession of the fascines and ladders, in conformity "with the said orders, knowing the period of attack "to be momently approaching." Lieutenant Colonel Mullins was found guilty of the above, and cashiered, but of that portion of another charge which implied want of personal courage, he was honourably acquitted. Indeed it would be hard to reconcile that charge with the tribute of commendation this officer had received on more than one occasion during the campaign, as specified in this record.

SERVICES OF THE SECOND BATTALION.

1803 UNDER the circumstances stated at page 38, a second battalion was added to the 44th Regiment, consisting of men raised in certain counties in Ireland for limited service, under the "*Army of Reserve Act*," passed in July, 1803, and the "*Additional Force Act*," passed on the 14th July, 1804, and was placed on the establishment of the army from the 9th of July, 1803. Its establishment was fixed at ten companies, consisting of fifty-four sergeants, twenty-two drummers, fifty corporals, and nine hundred and fifty privates.

The following officers were appointed to the second battalion :—

Lieut.-Colonel.—Thomas Nicoll.

Majors. { Henry Powlett.
Thomas Danser.

Captains.

Charles Madden.	John A. Johnson.
Henry Nixon.	Archibald McAuly.
Thomas Murphy.	John Cruice.
Marcus John Annesly.	Henry Debbieg.
Hon. Thos. Mullins.	J. Chilton L. Carter.

Lieutenants.

John Wynne.	James Ponsonby.
Thomas O'Reilly.	Alexander Andrews.
Thos. A. Ballard.	William Langley.
T. A. Dudie.	James Clegg Kennedy.
William Davis.	Arthur Newport.

Ensigns.

—— Sanders.	James Hardy Eustace.
George Rea.	John Goldrisk.
Roland Davis.	Albert Friend.
George Clarges Hill.	William Alexr. Craig.
John Blakeney.	

Adjutant.—Colin Meekisson.
Quartermaster.—Henry Jones.
Surgeon.—Robert Miller.
Assistant Surgeon.—William Vassall.

1804 The battalion embarked at Mallow on the 25th of March, 1804, and arrived at the Isle of Wight early in April. Here it remained until the 23rd of October, when it embarked at Cowes for Guernsey, and arrived at that island on the 25th of that month.

1805 to 1809 The Battalion remained in Guernsey* until August, 1809, when it proceeded to the neighbouring island of Alderney.

1810 On the 20th of March, 1810, the battalion embarked for Cadiz, which city was at that period besieged by a powerful French army under Marshal Soult. It arrived there on the 4th April, and was employed in erecting batteries and strengthening the defences of the place. Matagorda, a small fort across the harbour of Cadiz, and in front of the French lines, although frequently cannonaded, was held for fifty-five days by a garrison of three hundred men under Captain (the late General Sir Archibald) Maclaine of the old 94th, or Scots Brigade. The 44th, on arrival at Cadiz supplied a reinforcement for this fort under Lieutenants

* Colonel Burney states that the handsome Scotch snuff-mull, still in possession of the regiment, was presented, in 1807, by Captain David Fyffe, on his being transferred to the first battalion, while the second battalion was quartered in Guernsey; and that he recollects it quite new on the mess-table, when he joined the second battalion at that station in 1808.

1810 William Manby and Ralph John Twinberrow,* which assisted in its heroic defence. The toil was incessant and the men were employed day and night in filling sand-bags and repairing the breaches. The French were kept at bay: but as the fort became at last so demolished as to be untenable, the troops were withdrawn on the 22nd of April, and it was blown up.

The battalion remained at Cadiz and the Isla de Leon till the 24th of September, 1810, when it embarked for Portugal, and arrived at Lisbon on the 4th of October.

Now was about to commence the great contest for the possession of Portugal. Lieut.-General Viscount Wellington concentrated his forces, and occupied the strong position of Torres Vedras, about thirty miles from Lisbon. Here the British Commander showed a resistance which compelled Marshal Massena to relinquish his design upon Lisbon, notwithstanding his superior numbers, and the vain boast, that, in pursuance of the Imperial orders, he would drive the English leopards into the sea and plant the eagles of France on the towers of that city.

The 44th, under Lieut-Colonel Charles Bulkeley Egerton, advanced up the country, and towards the end of December joined the army in the lines of Torres Vedras, and was brigaded with the first battalion of the 4th and second battalion of the 30th Regiment, under the command of Major-General Dunlop, and in the fifth division,† under Major-General James Leith.

---

* Lieutenant Twinberrow died in the regiment at Calcutta shortly after its arrival in India.

† The fifth division was thus composed :—

First brigade, under Major-General Hay—1st (3rd battalion), 9th, and 38th (2nd battalion) Regiments.

Second brigade, under Major-General Dunlop—4th, 30th (2nd battalion), and 44th (2nd battalion) Regiments.

Marshal Massena retreated from the lines of Torres Vedras toward Spain in the beginning of March, 1811, and the British troops moving forward in pursuit, the enemy's rear was attacked and harassed with varied success. 1811.

Lieutenant Pearce, on the 30th of March, led a subdivision of the 44th light company over the bridge of Sabugal, which was commanded by two of the enemy's guns, and turned the right of General Regnier's division, then engaged with the light division. The whole of the brigaded light companies of the division crossed the Coa, in extended order, at various fordable places, and surprised the French, who were halted cooking their dinners, thus enabling the British to regale themselves actually on their enemy's soup.

On the morning of the 3rd of April the British troops were put in motion to turn the enemy's left on the Coa, above Sabugal, and to force the passage of the bridge of that town. In the action which ensued, when the French were making arrangements for a renewed attack upon the light division, the head of Major-General Dunlop's column crossed the bridge, and ascended the heights on the right flank of the enemy; the cavalry and third division also appearing, the French retired across the hills towards Rendo, after having sustained a heavy loss in killed and prisoners.

The French after the action of Sabugal, fell back upon Alfayates, and in five days the last of them crossed the frontier. The British thus stood triumphant on the confines of Portugal. The battalion formed part of the advance into Spain; and when Marshal Massena proceeded to the relief of Almeida, the allied army went into position to oppose his progress, but the fifth division taking post near Fort Conception, on the left of the line, was not actively engaged in the action at Fuentes d'Onor* on the

* While on the frontiers of Spain, a little before this battle, there was an alarm in the night, when Lieut.-Colonel Egerton, of the 44th, who then commanded the brigade, observed a man coming with a lantern, and of course called to him to "put that

1811 3rd of May, nor on the 5th of the same month, with the exception of its light companies. Captain Jessop commanded the light company of the 44th; Lieutenant William Pearce was slightly wounded, but did not quit the field, and four rank and file were also wounded.

After the repulse of the French, the garrison of Almeida despairing of assistance, the commandant made preparations for destroying the works and vacating the place. At midnight, on the 10th of May he exploded the mines, and sallying forth in a compact column, broke through the blockading force; he was followed, however, by a few men collected on the instant, and by the 36th Regiment. Early on the following morning the rear of the column was overtaken in the act of descending the deep chasm of Barba del Puerco, where many of the French were killed and wounded, and three hundred taken prisoners; the remainder escaped.

The light companies of the 30th and 44th were called out at midnight for this affair, and hurried off in pursuit of the escaping garrison. The bridge being narrow, these two companies were in time to charge a remnant of the fugitives who could not get over for the crush, and drove those

---

light out," exclaiming, "who can that unsoldierlike man be?" This was no other than Paymaster Williams, who had been a lawyer; and it appeared that, having issued three months' pay to the regiment a few days before, without obtaining signatures, and anticipating some casualties, he had brought his book to obtain the necessary vouchers from the officers. For this purpose, he presented it to Colonel Egerton, leading his mare over his shoulder. Having pointed out the irregularity of wanting signatures on going into action, he commenced writing, when the young officers, delighting in a practical joke, struck some fuzees under the tail of the Paymaster's horse, causing it to kick up, the book to fall, and the bottle of ink to be spilt over it. The Colonel enjoyed the fun, although he put a grave face on the proceedings, and the Paymaster lost his chance of signatures. Upon the reduction of the second battalion, the Paymaster, who said he carried all his accounts in his head, not having made any entries for two years, was kept without half-pay until he had completed his accounts, to do which he had to obtain signatures from officers dispersed all over France and England.

who would not surrender over the precipice, where many 1811
remained clinging to the rocks, while others fell into the stream below. The French commander asked permission to rescue these poor fellows from their perilous situation, which was granted by Captain Jessop, of the 44th, the senior officer on the spot; a party of the 31st Voltigeurs came over the bridge for the purpose, and they were recognized by the men of the 44th, as the same they had been engaged with on the 5th of May.

On the 15th of June, the battalion was stationed at Nava de Aver, in Portugal, where it remained three weeks forming part of the four divisions of infantry left behind the Agueda to observe the movements of the French army under Marshal Marmont during the time the second siege of Badajoz was undertaken. When the enemy moved southwards, the 44th directed its march by Castello Branco for the Alemtejo, and went into position on the Cayo, being hutted near the wood and town of Arronches until the 2nd of July, when it proceeded to Portalegre.

The battalion left its quarters in Portalegre on the 21st of July, and was hutted near that town for seven days, after which it proceeded to the camp at Quinta Lameira. Lord Wellington next moved towards the Agueda, with the view of rescuing Ciudad Rodrigo from the power of the enemy, and the 44th crossed the Tagus by the bridge of boats, and proceeding by Castello Branco entered Spain at Navas Frias, on the 11th of August. The battalion was stationed at Villa Rubias until the end of September; meanwhile, Ciudad Rodrigo was blockaded, and Lord Wellington's head-quarters were at Fuentes Guinaldo. When Marshal Marmont advanced to relieve this fortress, the battalion remained with its division at St. Payo, watching the passes from Estremadura. In the subsequent manœuvres it was not brought into action and after the retreat of the French army the battalion was stationed at Guarda, in Portugal. In November, the 44th, commanded by Lieut.-Colonel

1811 the Honourable George Carleton, was stationed at Villa de Rey, and in December at Oliveirinha.

1812 Being deceived by the seemingly careless winter attitude of the allied army, the enemy left *Ciudad Rodrigo* to the protection of its garrison, and the British commander, profiting by this omission, commenced the siege of that important fortress in the early part of January, 1812, with four divisions. On the receipt of intelligence that Marshal Marmont was collecting his forces to succour the place, the whole army was brought forward and posted in the villages on the Coa, ready to cross the Agueda and give battle. *Ciudad Rodrigo* was captured by storm during the night of the 19th of January, and the second battalion of the 44th was cantoned in the suburbs of that fortress from the 20th of January to the 23rd of February.

The fifth division afterwards proceeded towards the Alemtejo to engage in the siege of *Badajoz*, and the battalion joined its brigade at Campo Mayor on the 24th of March. On the 30th of that month it proceeded to Elvas; on the 4th of April it bivouacked near the city of *Badajoz*, and three breaches, which were deemed practicable, having been made, the battalion was formed on the evening of the 6th, ready to take part in the storming of this strong fortress. On this occasion the fifth division was directed to make a false attack on the Pardaleros, and a real assault on the bastion of San Vincente; the 44th, under Lieut.-Colonel the Honourable George Carleton, nobly distinguished itself on this occasion, and gained for the regimental colour an additional inscription.

At eight o'clock in the evening, the fifth division, under Lieut.-General Leith, of which the 44th formed part, broke up from its bivouac, and the second brigade, composed of the 4th, 30th and 44th, and the 5th Caçadores, commanded by Major-General Walker, formed in rear of a mill close to the south-west angle of the works, at which point it escaladed, and having succeeded in gaining the walls, maintained its ground, carried the bastions to the right, obtained possession of the Las Palmas gate, and cleared that part of the town of the enemy. 1812

Lieut.-Colonel Francis Brooke, of the 4th Regiment commanded the advanced party, which consisted of the 4th King's Own, the light companies of the 30th and 44th, and that of the 38th from the first brigade of the division. After breaking through the palisading and entering the ditch they were exposed to a murderous flanking fire of grape and musketry, and half their numbers fell without being able from their position to return a shot. Finding it impossible to fix a ladder, Lieut.-Colonel Brooke called for a volunteer to return to the reserve and report to Major-General Walker the state of affairs. Lieutenant Pearce, of the 44th, undertook this duty, and brought up two companies of the regiment under Captain John Cleland Guthrie,* who extended them on the glacis, and in a quarter of an hour silenced the two guns in the flank of the bastion and the musketry on the ramparts. The ladders were then quickly raised, and the stormers entered, followed by the brigade, the colours of the 44th being the first planted on the bastion of the fortress.

Soon after a footing had been established on the ramparts, a bugler of the 44th, about eighteen years of age, named Muchian, sounded the advance; Lord Wellington, who, with his staff, was near the two breaches where the attacks had hitherto failed, and listening with great anxiety, exclaimed, "There's an English bugle in the tower." This was the one in question. The British commander had just before remarked, if the attack "fails to-night I will take the tower in the morning "with these two regiments," meaning the 1st and 9th Foot then in reserve near him.

---

* Private John Collett, an old relic of Captain Guthrie's company, died at Cheltenham early in 1863. He was a Waterloo veteran, and had the Peninsular war-medal with five clasps, and ninepence a day pension.

1812 Captain Jervoise, Lieutenants Unthank and Pearce, three sergeants, and fifty rank and file composed the light company of the 44th. On Lieutenant Pearce calling over the roll in camp the next morning, there were left himself, Sergeant Jenkins and eighteen men. At daylight this officer found Lieutenant Unthank in the embrasure by which the rampart was entered, dying of a bullet wound in the side. The Chaplain of the division came up just in time to administer the Sacrament to him, which he received, supported on Lieutenant Pearce's knee. He was buried by his brother officer on the spot where he fell. Captain Jervoise died of his wounds on the 10th of July.

The castle having been taken, and the escalade of the bastion St. Vincente, by the fifth division, having turned the retrenchments, there was no further resistance, and the 4th and light divisions, whose heroic assaults on this occasion are matters of history, marched into the town by the breaches. The third division remained under arms in the castle until daylight, when the gate was opened, and permission given to enter the town.

In the Earl of Wellington's despatch the brigade was specially noticed, and it was therein stated that "the gallantry and conduct of Major-General Walker, "who was also wounded, and that of the officers and "troops under his command, were highly conspicuous." Lieut.-Colonel the Honourable George Carleton, who commanded the battalion, and obtained a medal, was also spoken of in terms of commendation.

The 44th had Lieutenants William Samuel Unthank and Matthew Argent, two sergeants, and thirty-five rank and file killed; Lieut.-Colonel the Honourable George Carleton had his jaw broken by a bullet; Captains John Berwick, Adam Brugh, and Francis B. Jervoise (died of wounds) were wounded severely; Lieutenant William Henry Meade slightly; Lieutenant Temple Frederick Sinclair severely; and Ensign John O'Reilly slightly. Seven sergeants, one drummer, and eighty rank and file wounded. Total casualties, one hundred and thirty-four. 1812

This storming of *Badajoz*, in which the 44th bore so distinguished a part, was probably the most celebrated of the many gallant exploits performed by the Peninsular army, and is thus spoken of by one of the highest authorities on the subject.

Sir John Jones, in his "Journal of Sieges in Spain," says, "In ordinary military reasoning, such a spot would be "considered secure from assault, but the efforts of British "troops occasionally set all calculation at defiance; and when "a few years shall have swept away the eye-witnesses of their "achievements on this night, they will not be credited."

The word "BADAJOZ," on the Regimental Colour, granted by royal authority, commemorates the services of the 44th on this occasion.

Shortly after the capture of Badajoz the 44th marched with the army towards the Agueda; on the 14th of May the second battalion of the 4th Foot joined from Ceuta, and the brigade, then consisting of the two battalions of the 4th, and second battalions of the 30th and 44th, was commanded by Major-General Pringle, Major-General Walker having returned to England in consequence of his wounds.

The 44th advanced with the army to Salamanca, and this city being rescued from the power of the enemy, exhibited a scene of tumultuous rejoicing; the houses were illuminated, and the people, shouting and singing, welcomed their deliverers; the army took up a position on the mountain of St. Christoval, a few miles in advance, but the 44th remained behind the Tormes. The forts which the French occupied at Salamanca were besieged, and when Marshal Marmont advanced on the 20th of June to their succour, the battalion crossed the Tormes, and was formed with the army in order of battle on the top of the mountain. After the capture of the forts, the enemy retired towards the Douro, and the battalion descended the heights and followed the French army, which took up a new position near Tordesillas.

In front of this place the 44th bivouacked from the 2nd

1812 to the 9th of July, and at Nava del Rey until the 16th; the weather being fine, the country rich, rations regularly supplied, and wine abundant, the soldiers fared luxuriously; but the enemy having passed the Douro and the Trebancos, turned the left of the allies, and marched on the 18th of July towards the Guarena, when the British retired. The two armies directed their march towards the Guarena as to one common goal, and important results depending on which should first pass the stream, a trial of speed arose. Several of the hostile columns proceeded a distance of ten miles in parallel lines within musket-shot of each other, and marched impetuously towards the stream in perfect order amidst clouds of dust. A most interesting spectacle presented itself; for the officers on each side, being strangers alike to malice and to fear, were seen pointing with their swords, touching their caps, or waving their hands in courtesy, as they urged their men in their course towards the river. The British gained the stream first; the soldiers being tormented with thirst, many of them drank as they marched, and others halting in the river a few moments, were saluted with a shower of bullets; but they passed on, and the French marshal's designs were frustrated. He, however, crossed the river on the 20th July higher up, turned the right flank of the allied army, and gained a new range of hills; Lord Wellington made a corresponding movement, and an evolution similar to that on the 18th was repeated, which ended in the British resuming their position on the heights of St. Christoval. The 44th forded the Tormes on the following day, and was posted in position with the army near *Salamanca*.

These bold manœuvres of the enemy were watched by the British commander, who waited patiently for an opportunity to strike a decisive blow; and this occurring on the 22nd of July, a sanguinary battle was fought. In the early part of that day the 44th was posted on the slope of of one of the heights, called the Arapiles, where it remained until the afternoon, when it moved to the rear of the village of that name; Lord Wellington having detected a fault in his adversaries' movements, ordered his divisions forward, and the battle commenced.

In this action the fifth division, of which the 44th formed part, under Lieut.-General James Leith,* attacking the enemy in front, highly distinguished itself, and the battalion, commanded by Brevet Lieut.-Colonel George Hardinge, had a prominent share in the glories of the day. 1812

This division advanced in two lines, left in front, which brought the second brigade in front, and the 44th on the right of the first line; the light companies of the brigade (4th, 30th, and 44th) being in skirmishing order, keeping up the communication between the fifth division and the third division on its right. As the division advanced the light companies closed to their left and joined the right of the first line (or 44th Regiment) on the east of the hill, which was so thickly covered with flax and evergreen oak that the division and the enemy came, as it were, unexpectedly into each other's presence, at no greater distance than from fifteen to twenty yards. The French were at this moment deploying from grand divisions, and the second grand division had but half of its front clear, when, by a simultaneous impulse, both parties fired a volley: the British instantly charged, and the French, being in confusion, turned. Two Squadrons of the 5th Dragoon Guards went right through them on their flank, and the whole of them, with their eagle and two standards, were captured. The eagle,† which was that of the 62nd Regiment of the line was taken by Lieutenant W. Pearce,‡

* Then serving with the local rank of Lieut.-General in the Peninsula; he was made a K.B. on 1st February, 1813, a Lieut.-General on the 4th of June following, and a G.C.B. on the 2nd January, 1815; he died on the 16th of October, 1816.

† Another trophy of this action, long preserved in the 44th Regiment, was a drum, taken from the French battalion they were engaged with. It had got rather beyond use, and was left with the depôt company when the regiment embarked for the Mediterranean in 1848. In the numerous changes of the station of the regimental depôt, the "Salamanca drum" was unfortunately lost.

‡ This veteran officer, then Lieut.-Colonel Pearce, K.H., and in his seventy-fifth year, furnished some very interesting details relating to the 44th Regiment, for the first edition of this record: he received the Peninsula war-medal with 3 clasps, for Fuentes d'Onor, Badajoz, and Salamanca. For the capture of the eagle, he obtained,

1812 44th Regiment, and the two standards by Lieut. Francis Maguire, 4th, and Ensign John Pratt, 30th Regiment. The French were called on to lay down their arms, about which there was some demur; but on the cavalry returning to the charge they gladly did so, craving protection. Ensign Standley was killed carrying one of the colours of the 44th.

The French officer who carried the eagle had just wrenched it from the pole, and when Lieutenant Pearce first saw it, he was endeavouring to conceal it under the grey great-coat, which he wore over his uniform; Private (afterwards Sergeant) Finley aided in the capture, and the French officer making resistance, was assisted by one of his men, who attacking Lieutenant Pearce with his fixed bayonet, was shot dead by Private Bill Murray, of the 44th light company. Privates Blackburn and Devine, of the same company, had also a hand in this affair, and Lieutenant Pearce divided twenty dollars—all the money he had with him—amongst the four, for their gallant exertions.

Soon after the capture of the eagle a heavy column of French infantry came in sight, although at a considerable distance; the eagle was at once placed on a sergeant's halbert, the men giving three cheers.

This trophy was kept for the night with the regimental

---

in August of the following year, a company in the seventh battalion of the 60th, composed of Germans of the Contingent of the Rhine, who had deserted to the British in a body. In order to mark its identity, Lieutenant Pearce scratched his name on the pedestal with a nail. Some years after, he went to see the eagles which had been removed from Whitehall Chapel to Chelsea College. Lieut.-Colonel Le Blanc, then Major of the Hospital, told him there was one that could not be accounted for, which he recognised as that captured by himself at Salamanca. Since the period of this visit it has been ascertained that one was missing, and it is considered it must be the eagle in question. Captain Ford, Captain of Invalids, Chelsea Hospital, in a paper published in "Colburn's United Service Journal" for April, 1844, states that the two eagles taken at Salamanca were then in the Chapel of the Royal Hospital at Chelsea. It is singular that the late Lord Clyde was also recommended for his company by Lord Wellington, which he obtained in the *eighth* battalion of the 60th, in the same year, he having also been in the *fifth* division, his regiment being the 9th Foot.

quarter guard, and Lieut.-Colonel Hardinge sent to Major-General Pringle on the following morning, to know what was to be done with the eagle. "Send it to the man who took it," was the reply: and Lieutenant Pearce carried it on the march the next day and night, and delivered it the following day, the first time a halt was made, at head-quarters, at a village from which Lord Wellington wrote his despatch about the battle. 1812

His lordship's skill was bravely seconded by the resolute valour and discipline of the troops. The action lasted until dark. A decisive victory was gained over the French, and eleven pieces of cannon, two eagles,* and six colours were captured. Lieut.-General Leith was wounded during the action, and Major-General Pringle assumed the command of the fifth division.

In his despatch, the Earl of Wellington observed, that "in a case in wh ch the conduct of all has been conspicuously "good, I regret that the necessary limits of a despatch prevent "me from drawing your Lordship's notice to the conduct of a "large number of individuals; but I can assure your Lordship "that there was no officer or corps engaged in this action "who did not perform his duty by his Sovereign and his "Country."

Brevet Lieut.-Colonel Hardinge gained a medal for this victory, and the 44th were permitted to bear the word "*Salamanca*" on the regimental colour and appointments. The battalion had Captain John Berwick, Ensign William Standley, and four rank and file killed; two sergeants, one drummer, and twenty rank and file wounded.

The troops pursued the flying enemy on the following day, and after numerous marches and manœuvres the second battalion of the 44th proceeded with the army to Madrid, where the British were received with acclamations of joy. There being no ensigns in the battalion, the King's colour

---

* The other eagle taken at Salamanca was captured by a Cacadore battalion, commanded by Major Crookshank of the 38th, which latter regiment was in the sar- division (the fifth) as the 44th.

1812 was carried through the city by Lieutenant Pearce. In August the battalion was stationed at Villa Castine, and afterwards proceeded to the camp near *Burgos*, during the siege of that Castle, where it arrived in September.

When the concentration of the enemy's forces and the failure of the attack on Burgos rendered a retrograde movement necessary, the army withdrew in the night, and the French following in full career, several skirmishes took place. On the 24th of October the army was in position behind the *Carion*, and the 44th bivouacked near *Villa Muriel*. The enemy attacked the left of the British army at *Villa Muriel** on the 25th of that month, but were repulsed by the fifth division of infantry under Major-General Oswald, Lieut.-General Leith being absent, on account of indisposition. The bridge of Villa Muriel was destroyed, but the enemy discovering a ford, passed over a considerable body of infantry and cavalry. The French troops were driven back, the village was re-occupied in force, and the canal was lined by the allied troops ; but the army withdrew before daybreak on the 26th, and the retreat was resumed. In this action the 44th was engaged from daybreak until midnight, and its successful efforts enabled the army to effect its retreat. It had the following officers killed and wounded; Lieutenant William Fitzmaurice Lennon killed, and Lieutenant Henry Elwin and Ensign Michael Smith wounded ; these died of their wounds, the former on the 14th November, and the latter on the 30th October, 1812. The strength of the battalion was reduced to only forty-two men of all ranks fit

* On the day before this action, while the French were pressing the British hard with their cavalry, the printing press of the Spanish General Castanos stuck in passing a small stream, and the light companies, on rear guard, were compelled to halt to rescue it. The Marquis of Wellington coming to see what was the matter, up jumped a hare, when a brace of greyhounds belonging to Lieutenant Robert Grier, of the 44th, happening to be at hand, started off, followed by his Lordship and an orderly. It was a pretty course, and seen by all present. Wellington made the orderly dismount, and sent back the dead hare. The dogs were afterwards brought to England. Lieutenant Pearce had a Welsh spaniel named Brush that would run up and down the whole division, frequently flushing game to the amusement of the men. Poor Brush died on service.

for duty at regimental head-quarters.* All the wounded 1812
were left on the ground, the baggage and women of the fifth division falling into the hands of the enemy. Assistant-Surgeon Collins of the 44th was sent on the following morning to take care of the wounded, and had a rouleau of doubloons given to him by the Commissary-General; a drunken trumpeter of the French army, however, was the first man that met him after starting, and robbed him of the money. The poor doctor was then marched to the rear, and never saw the ill-fated wounded soldiers, who were left to the mercies of the enemy's surgeons.

In a retreat like that from Burgos, the sufferings of an army—such as the loss of baggage and stores, the wounded and women falling into the hands of the enemy, the frequent attacks from overwhelming numbers, and not least the difficulty of procuring rations are necessarily great, more especially to the division covering the retreat. At one time the men were without biscuit for eleven days, and received only a small ration of beef, and that of the leanest and toughest kind, for all the cattle they could get travelled with the army. The survivors were given each a pair of boots as compensation. Nor were the officers better off than the men, as some were seen marching barefooted and bleeding, their baggage mules (one between two officers) having broken down, and no one able, if willing, to spare anything from his own small kit.

The army took up positions from time to time to retard the advance of the enemy, and having crossed the Douro and the Tormes, it was posted behind the latter river until the middle of November, when a further retreat to Ciudad Rodrigo took place. At this period, the battalion was stationed at Castellegos, and in December at St. Martinha.

On the 6th of December, 1812, the Marquis of Wellington,

---

* Sergeant Farrell, of the seventh company, came up the following morning to Lieutenant Pearce, who, although attached to the light company throughout, had the paying of the seventh, and said, "Sir, the mule and camp-kettles are lost, but as I am the only "man of the company left, it is not of much consequence."

1812 which title was conferred upon him after the victory of Salamanca, issued an order desiring that the men, fit for service, of the second battalion of the 44th, which was so reduced in numbers, should be formed into four companies, and the remaining six sent to England. The battalion at this period (a draft of forty-four non-commissioned officers and men having recently arrived from England) consisted of thirty-six sergeants, twenty drummers and three hundred and ninety-nine rank and file, but only one hundred and thirty of the latter were reported "fit for duty," the remainder, with the exception of nineteen "on command," being sick or wounded.

1813 Accordingly, six companies, consisting of six officers, twelve sergeants, twelve drummers, and fifty-four rank and file, embarked at Lisbon on the 13th of February, 1813, and arrived at Portsmouth on the 24th of that month. The four companies remaining on service in the Peninsula, consisting of twelve officers, nineteen sergeants, ten drummers, and three hundred and eleven rank and file were incorporated with four companies of the second battalion of the 30th Regiment, similarly left, and together numbered the fourth provisional battalion. These two corps being thrown so much together in Spain, were almost like one regiment. During its service in that country the second battalion of the 44th was known as the "Little Fighting Fours," from its being a hard-fighting corps, and the men of small stature.

In the summer the second battalion of the 59th Regiment having joined the fifth division, it relieved the provisional battalion, which proceeded to England. The four companies embarked at Lisbon on the 15th of June, and arrived at Portsmouth on the 13th of July.

The second battalion was subsequently authorized to bear the word "PENINSULA" on its colour and appointments. It was not destined, however, to remain any length of time at home, for in the short space of about four months, and after recruiting its ranks, the battalion embarked for Holland, as part of the force under Lieut.-General Sir Thomas Graham, K.B. During its short stay in England, the battalion was stationed at Horsham, and afterwards at Brighton and

Steyning: here the battalion furnished the guard for the 1813
Prince Regent. It embarked at Ramsgate on the 29th of November, 1813, mustering about five hundred men, under the command of Lieut.-Colonel the Hon. George Carleton, and landed at Tholen, in Zeeland, on the 17th of December, 1813. The first brigade, under Major-General Skerrett, was composed of the second battalions of the 37th, 44th, and 69th Regiments.

A forward movement upon *Antwerp*, at the request of 1814
the Prussian General, Bulow, was made to favour his operations, and that place was accordingly bombarded by the British forces on the 13th of January, 1814.

The winter was intensely cold, and the duties in the trenches before Antwerp were severely felt. Lieut.-General Sir Thomas Graham (afterwards Lord Lynedoch), on making a reconnaissance of that place, was attacked by a body of French troops at *Merxem*, on the 13th of January, but the battalion (though employed on this duty) did not suffer any loss on the occasion. An attack on the village of *Merxem*, where the enemy was strongly posted, was made on the 2nd of February, but the battalion sustained no casualties. Afterwards the British troops were employed in constructing a breastwork and battery. On the 3rd of February, several pieces of heavy ordnance opened upon the city of Antwerp, and on the French shipping in the Scheldt; the cannonade was continued until the 6th, when General Bulow having received orders to march southward to act with the grand army of the allies, it became necessary to relinquish the attack on Antwerp, and the British retired towards Breda. In the list of casualties from the 3rd to the 5th of February, Ensign Alexander Reddock, of the 44th was reported as slightly wounded. The second battalion was stationed near Wouw, until about the end of February.

In the beginning of March the battalion again moved towards Antwerp, and was employed on the night of the 8th of that month in the attempt to carry by a *coup-de-main*, the strong fortress of *Bergen-op-Zoom*. The battalion formed part of the fourth column, commanded by Brigadier-General

1814 Gore, and Lieut.-Colonel the Hon. George Carleton,* of the 44th. The column consisted of the 44th, three hundred; flank companies of the 21st and 37th, two hundred; Royals, six hundred; in all, one thousand one hundred men. Major-General Skerrett and Brigadier-General Gore both accompanied the fourth and right column, which was the first to force its way into the body of the place. The two columns told off for the assault were directed to move along the ramparts, so as to form a junction as soon as possible, and having cleared them, then to assist the centre column, or to force open the Antwerp Gate. An unexpected difficulty about passing the ditch on the ice having obliged Major-General Cooke to change the point of attack, a considerable delay ensued, and the left column did not gain the rampart until half-past eleven. Meanwhile, the fall of Brigadier-General Gore and Lieut.-Colonel the Hon. George Carleton, and the dangerous wound of Major-General Skerrett, depriving the right column of its proper leaders, it fell into disorder, and suffered great loss in killed, wounded, and prisoners. The centre column, which was driven back, formed again, and advanced to effect a junction with the left column on the ramparts. At daylight the besieged turned the guns on the British, who were without protection on the outworks, and the officers and men that had penetrated the works were forced to surrender prisoners of war; this measure was adopted by Major-General Cooke as, there being no prospect of relief, a further continuance of the contest would only have caused a fruitless waste of life.

---

* When the regiment was in Holland, Lieut.-Colonel Carleton, as a compliment, had two Cossacks attached to him as orderlies. On one occasion they were billeted with Captain Power, of the regiment, who, of course, was entitled to the best room. To secure this for his own family, the master of the house took to his bed apparently sick. These Cossacks were eating, drinking, and smoking all day: and in the evening one of them went with a large wisp of lighted straw to feed the horses. Some of the inmates ran to the patient in alarm, exclaiming that the foreigners were burning the house down. The sick man thereupon jumped out of bed to see what was the matter, upon which the other Cossack, with spurs, boots, and clothes, immediately took the unoccupied place, thus securing it for the night for himself and comrade.

The 44th, with the right column, had succeeded in forcing its way into the town, and had occupied nine bastions of the fortress, its colours being the first displayed on the ramparts. Captain Burney, of the 44th, who acted as brigade-major to Lieut.-Colonel Hon. G. Carleton, was the first who scaled the wall. By the aid of two sergeants' pikes, the cross-bars of which were placed so as to give him a footing, he forced himself through the fraise, or palisading, at the salient angle of bastion (No. 15); one of the colours of the 44th was then thrown up to him, and displayed by him on the rampart. The column being unable to follow for want of ladders, he was ordered to throw down the colour again, and to assist the column in its entrance by a dyke, towards which the guides were leading; this dyke passed through an arched passage four feet high, beneath the rampart, and was the water communication between the ditch and the canal within the town; the iron bars at each end of the passage having been removed by the towns-people who were friendly to the British. Captain Burney accordingly went round by the rampart, the sentry at that point having made off, and meeting the men as they entered, formed up a party of different corps, and drove back the enemy's main guard. The column then forced its way along the ramparts towards the other points of attack, but being soon without a leader to direct it, owing to the fall of Generals Skerrett and Gore and Lieut.-Colonel Carleton,—Brevet Lieut.-Colonel Hardinge of the 44th being also wounded,—it separated into parties in the different bastions, the 44th in several, under various captains of the regiment, and turned the guns upon the town. The inhabitants were in favour of the British, yet owing to the want of nerve on the part of the guides, the columns of the other points of attack were behind their time: the garrison rallied, recovered possession of the gates, and repulsed the columns that had not yet entered, The parties still within the walls, attacked in their bastions. and fired on from the windows of the opposite houses, suffered severely, and at daybreak were ordered by Major-General Cooke to lay down their arms and surrender, permission being given to Captain Burney, with the colours, Captain Guthrie, and a few other officers and men of the

1814

1814 44th, to escape from the town, in consideration of their gallantry and services during the night.

The battalion suffered severely, Lieut.-Colonel the Honourable George Carleton and Ensign James Maxwell were killed; Major George Hardinge (Brevet Lieut.-Colonel) was severely, though not dangerously, wounded, and taken prisoner; Captain John Cleland Guthrie, slightly, Captains David Power and Thomas A. Ballard, severely, afterwards died of wounds (prisoners); Captain Thomas A. Dudie, severely; Lieutenant George Clark Beatty, slightly (prisoner); Lieutenants John O'Reilly and William Tomkins, severely (prisoners); Ensigns Henry Martin and Gillespie Dunlevie, severely: Adjutant William Henry Meade, died of wounds, and Ensign Benjamin Whitney, severely (prisoners). The latter officer was taken prisoner while planting the King's colour on the ramparts. Eleven sergeants, three drummers, and two hundred and five rank and file were taken prisoners, as also Captain George Crosier, Lieutenants Frederick Hemming, Ralph John Twinberrow, and Temple Frederick Sinclair. The return of killed and wounded of the 44th has not been preserved separately from the whole force employed; but in the monthly return, however, of the battalion of the 25th of March, 1814, two sergeants, one drummer, and thirty-eight rank and file are reported as "*dead.*"

A soldier of the 44th, named McCullop, who had received nine hundred lashes within nine weeks, and on the night of the assault was a prisoner, begged to be released, saying he had never been out of fire when the regiment had been engaged since his joining, and although he knew that he was a bad soldier in quarters, yet he was a good one in the field. The man had his wish, and, being an excellent shot, managed to kill the first nine sentries that were met with:—he was killed, however, during the night.

On the 28th of March, 1814, the second battalions of the 21st, 36th and 44th, and the 55th Regiment, were formed into the second provisional battalion; and of the 44th, under Captain John Cleland Guthrie, mustered only twenty-one sergeants, thirteen drummers, and one hundred and seventy-two rank and file.

Peace was shortly afterward concluded. On the 4th of 1814
April, Napoleon Bonaparte signed his abdication in favour of his son; but this proposal being rejected, he subsequently renounced the throne of France and Italy for himself and heirs. Having selected Elba for his residence, that island was ceded to him in full sovereignty for life, with a pension payable from the revenues of France; and by the treaty which was signed at Paris on the 11th of April between the Allies and Napoleon, it was agreed that he should retain the imperial title during his lifetime. Ample pensions were also assigned to his relatives.

On the 3rd of May, 1814, Louis XVIII. entered his capital, and re-ascended the throne of France, and on the 30th of that month, the general peace between that country and the allied powers of Austria, Russia, Great Britain, and Prussia, was signed at Paris.

Meanwhile, the officers and men taken prisoners at Bergen-op-Zoom were released; the 44th composed part of the second brigade of the second division, consisting of the second battalions of the 25th and 73rd, and the 55th Regiment. In August, the 44th formed part of the third brigade, under Colonel Macleod, consisting of the second battalions of the 35th, 44th and 78th, and the 54th Regiment, with a detachment of the 95th, now Rifle Brigade.

The battalion occupied quarters in Ostend from May, 1815
1814, until April of the following year, when events occasioned it to be again employed on active service, and to acquire its crowning triumph. During its stay at Ostend the battalion acquired the good-will of the inhabitants, who, for years after spoke of the officers and men in very gratifying terms.*

Louis XVIII. at the beginning of the year 1815, seemed apparently firmly seated on the throne of France; but various

* While the 44th were at Ostend, a Prussian Regiment was quartered there, and the officers of the latter were invited to the mess of the British Regiment. The Prussians kept on their swords during dinner, and on being invited to lay them aside, declined, saying it was contrary to orders, as they did not know but that they might have to use them before quitting the room.

1815 causes of discontent, existed in that country. The army, long accustomed to war, still retained a chivalrous veneration for Napoleon Bonaparte, who was kept acquainted with the state of the public mind, and this feeling of his former troops. On the evening of the 26th of February, he embarked at Porto Ferrajo, in the island of Elba, with about a thousand troops, of whom a few were French, and the remainder Poles, Corsicans, Neapolitans, and Elbese. With this motley band he landed at Cannes, in Provence, on the 1st of March, 1815, and the result proved that his calculations were correct. After being joined by the garrison of Grenoble, he proceeded to Lyons, and entered that city amidst the acclamations of "*Vive l'Empereur!*" from the soldiers and the people. The possession of the second city in France being thus obtained, Napoleon reassumed the dignity of Emperor, and continued his advance to Paris, which he reached on the 20th of March, his progress having been one continued triumph.

In the meantime, Louis XVIII. had withdrawn from Paris to Ghent, and Napoleon regained possession of the throne of France as Emperor; however, the allied powers refused to acknowledge his sovereignty, and determined to effect his downfall.

On the 11th of April, 1815, it was announced to the army in Flanders, that His Royal Highness the Prince Regent, in the name and behalf of His Majesty, had appointed Field-Marshal the Duke of Wellington, K.G., to be commander of His Majesty's forces on the continent of Europe; and it was directed that the ninth British brigade of infantry should be composed of the third battalion of the 1st Foot, the 42nd, second battalion of the 44th, and the 92nd Regiment, under the command of Major-General Sir Denis Pack, K.C.B.

It had been a busy time for the 44th whilst stationed at Ostend, as, to save tide, it furnished fatigue parties night and day for landing horses, artillery, and stores of every description for an army in the field. The transports came in six at a time with one tide, and departed by the next. As the tide left them dry on the sands, every thing that could not spoil was turned overboard, and the 44th fatigue parties carried them up out of the way of the next tide. The transport being then light, was pulled over to the wooden jetty on the other side of the harbour, where the fragile stores which would have spoiled on the damp sands were landed. The officers of the 44th were in billets at Ostend, but the men occupied a barrack near the rampart facing the sea. The battalion, on the 25th of May, consisted of thirty-two officers, forty-four sergeants, sixteen drummers, and four hundred and fifty-five rank and file, and was commanded by Lieut.-Colonel Hamerton. 1815

Napoleon left Paris on the 12th of June, and endeavoured, by one of those rapid and decisive movements for which he had been so celebrated, to interpose his forces between the British and Prussian armies, and then attack them in detail. Information of this movement arrived at Brussels during the evening of the 15th of June, and the troops were immediately ordered to prepare to march.

The fifth division*—under Lieut.-General Sir Thomas Picton—marched at daylight, on the 16th of June, and at about half-past two o'clock came within range of the enemy's artillery at some farm-houses denominated *Les Quatre Bras*, in front of Genappe, where the main road from Charleroi to Brussels is crossed by another from Nivelles to Namur, and which latter served as the British communication with the Prussians on the left.

As the British regiments arrived on the scene of action, they were instantly placed in position. The repeated charges of the French were repulsed, but a considerable loss was incurred, including his Serene Highness the Duke of Brunswick, who fell at the head of his troops.

Captain Siborne—in his detailed "History of the War in France and Belgium in 1815"—has depicted in glowing

* The fifth division consisted of a brigade commanded by Major-General Sir James Kempt, K.C.B.—28th, 32nd, 79th, and 95th (Rifles) Regiments—numbered fifth in reference to the whole of the infantry; and a brigade commanded by Major-General Sir Denis Pack, K.C.B.—1st (third battalion), 42nd, 44th (second battalion), and 92nd Regiments—numbered ninth; and an Hanoverian brigade.

1815 and eloquent terms the attack of the French cavalry on the 42nd Highlanders, and 44th Regiment—as will be seen by the following extract:—

"If this cavalry attack had fallen so unexpectedly upon "the 42nd Highlanders, still less had it been anticipated by "the 44th Regiment.* Lieut.-Colonel Hamerton—perceiving "that the lancers were rapidly advancing against his rear, "and that any attempt to form square would be attended "with imminent danger, instantly decided upon receiving "them in line.† The low thundering sound of their approach "was heard by his men before a conviction they were French "flashed across the minds of any but the *old* soldiers who "had previously fired at them as they passed their flank. "Hamerton's words of command were:—'Rear rank, right "about face!'—'Make ready!' (a short pause, to admit of the "still nearer approach of the cavalry)—'Present!'—'Fire!' "The effect produced by this volley was astonishing. The "men—aware of their perilous position—doubtless took most "deliberate aim at their opponents, who were thrown into "great confusion. Some few daring fellows made a dash at "the centre of the battalion, hoping to capture the colours, "in their apparently exposed situation; but the attempt, "though gallantly made, was as gallantly defeated. The "lancers now commenced a flight towards the French position "by the flanks of the 44th. As they rushed past the left "flank, the officer commanding the light company‡—who

* "Meanwhile Pack's brigade—consisting of the Royals, 42nd "44th, and 92nd, which here upheld their noble character— "succeeded, after an arduous conflict, in repulsing the enemy on "the left of the high road. The third of these regiments, (the "44th) being suddenly assailed by lancers in rear, when engaged in "front, and having no time to form square, performed the astonish- "ing feat of receiving the cavalry *in line*, and defeating it by a "single well-directed discharge of the rear rank, who faced about "for that purpose."—*Alison's History of Europe.*

† This is regarded as the first instance of so noble a defence against French cavalry, and their entire rout by a regiment of British infantry in line.

‡ Lieutenant Campbell commanded the light company on Capt. Brugh and Lieutenant Grier being wounded.

"had very judiciously restrained his men from joining in the 1815 "volley given to the rear—opened upon them a scattering "fire; and no sooner did the lancers appear in the proper "front of the regiment, when the front rank began in its "turn to contribute to their overthrow and destruction.

"Never, perhaps, did British infantry display its "characteristic coolness and steadiness more eminently than "on this trying occasion. To have stood in a thin two-deep "line, awaiting, and prepared to receive, the onset of hostile "cavalry, would have been looked upon at least as a most "hazardous experiment; but, with its rear so suddenly "menaced, and its flanks unsupported, to have instantly "faced only one rank about—to have stood as if rooted to "the ground—to have repulsed its assailants with so steady "and well-directed a fire that numbers of them were des-"troyed—this was a feat of arms which the oldest or best-"disciplined corps in the world might have in vain hoped to "accomplish; yet most successfully and completely was this "achieved by the gallant second battalion of the 44th "British Regiment, under its brave commander, Lieut.-"Colonel Hamerton.

"In this attack occurred one of those incidents which, "in daring, equal any of the feats of ancient chivalry; which "make the wildest fables of the deeds of the knights of old "appear almost possible; which cause the bearing of an "individual to stand out, as it were, in relief amidst the "operations of the masses; and which, by their characteristic "recklessness, almost invariably insure at least a partial "success. A French lancer gallantly charged at the colours, "and severely wounded Ensign Christie,* who carried one of "them, by a thrust of his lance, which, entering the left eye, "penetrated to the lower jaw. The Frenchman then "endeavoured to seize the standard; but the brave Christie, "notwithstanding the agony of his wound, with a presence "of mind almost unequalled, flung himself upon it—not to

* Promoted from Sergeant-Major of the 44th, on the 26th November, 1812; his gallanty at Quatre Bras gained him his lieutenancy in the regiment on the 26th October, 1815.

1815 "save himself, but to preserve the honour of his regiment. "As the colour fluttered in its fall, the Frenchman tore off "a portion of the silk with the point of his lance; but he "was not permitted to bear the fragment beyond the ranks. "Both shot and bayonetted by the nearest of the soldiers of "the 44th, he was borne to the earth, paying with the "sacrifice of his life for his display of unavailing bravery."

Captain Siborne, in a note on this circumstance, adds, "The part of the colour thus torn off by the French lancer, "is to this day in the possession of Major-General O'Malley, "C.B., then Lieut.-Colonel of the 44th Regiment,* to the "command of which he succeeded, on Lieut.-Colonel "Hamerton† being wounded at a later period of the battle. "The colours themselves are in the safe keeping of the latter "officer, whose decision, firmness, and bravery, on this "occasion, well entitle him to the guardianship of these "sacred and glorious relics of his gallant corps."

As the British regiments, on their arrival on the field, were assailed by enemies advancing from different points, each regiment had to fight independently, and, in many instances, to stand or fall by itself.

Captain Siborne, who had made the battle of Waterloo the study of his life, as is shown by his admirable model, now to be seen at the Royal United Service Museum, thus describes a subsequent attack of French cavalry at *Quatre Bras*:—

"It was not long before the British battalions most in "advance were warned of the approach of hostile cavalry "by the running in of their skirmishers; and scarcely had "they formed their squares, when (the batteries respectively

---

* Major-General George O'Malley, C.B., died on the 16th May, 1843. The part of the colour was subsequently presented to the regiment, and is still preserved as a sacred relic.

† Lieut.-General Hamerton was residing at Dublin when the regiment was quartered there in 1845. He most hospitably gave a ball to the officers of the corps, which he had led to such distinguished service. The glorious colours of the old second battalion were conspicuous amongst the ornaments of the room, and were quite the lions of the evening. He afterwards became General, and was appointed Colonel of the 55th Regiment. General Hamerton, C.B., died at Tipperary, on the 28th January, 1855.

"opposed to them having ceased their fire) a rushing sound 1815
"was heard through the tall corn, which, gradually bending, "disclosed to their view the heads of the attacking columns; "and now began a conflict wherein the cool and daring "intrepidity with which the British infantry are accustomed "to defy the assaults of cavalry, was exemplified in a "manner that will ever reflect honour and glory upon "the regiments to whose lot it fell, on this memorable field, "to assert and maintain their country's prowess. A rolling "fire from the muskets of the 42nd Highlanders and 44th "British regiment, given at a moment when the enemy's "horsemen were almost close upon their bayonets, though "most destructive in its effects upon their own immediate "opponents, checked not the ardour and impetuosity of the "general attack. These two diminutive squares, now "completely surrounded by the French cavalry, seemed "destined to become a sacrifice to the fury with which a rapid "succession of attacks was made upon them; no sooner was "one squadron hurled back in confusion than another rushed "impetuously forward upon the same face of a square, to "experience a similar fate; and sometimes different faces were "charged simultaneously. A strong body of cuirassiers now "passed the right flank of the two regiments, along the high "road, with an evident intention of making another attempt "upon Quatre Bras."

Sir Thomas Picton, seeing it was impossible to obtain support from the allied cavalry then in the field, and watching anxiously the contest maintained by the 42nd and 44th Regiments in their exposed situation, determined to aid the devoted squares, and as a substitute for cavalry assailed that of the enemy with some regiments of infantry. Accordingly, the third battalion of the Royals, and the first battalion of the 28th, advanced into the thick of the enemy's cavalry, formed squares, and the repeated charges were gallantly repulsed by them, supported by the 32nd Regiment.

The Duke of Wellington in his despatch, giving an account of the action at *Quatre Bras* thus alluded to the fifth division:—"The troops of the fifth division, and those of the

1815 "Brunswick corps were long and severely engaged, and "conducted themselves with the utmost gallantry."

Colonel Burney, K.H., then Captain Burney, 44th, gives the following account of the night of the 15th June and the day following:—"There was a grand ball in Brussels on the "15th June. I did not go for this reason: one of the "Quarter-Master-General's department was billetted in the "same house that I was: he rode in about four o'clock that "day, having been reconnoitring twenty or thirty miles in "front, with only one mounted man, and went straight to "the Duke of Wellington and delivered his observations. "On arriving at his billet, all was bustle. I found he had "ordered his servants to look well to his horses, and prepare "everything for a sudden move to the front. I took the "hint, packed my baggage and went early to bed. At "twelve o'clock that night, bugles were sounding and drums "were beating in every part of Brussels; people galloping "about in all directions; the ball broken up, and the troops "collecting at the alarm posts of their regiments. The orders "were to draw three days' provisions, and to cook on the "spot if time would admit before advancing. The 44th were "drawing rations and ammunition till four o'clock, when they "had to march out of Brussels, pretty well fagged, having "been accoutred since twelve. The regiment reached Quatre "Bras about one or two o'clock; halted and ordered to cook. "The Duke of Brunswick with his dragoons lay on one side of "the road, the 44th on the other: we could distinctly hear "artillery. The Prussians were engaged, and several of "their wounded passed us saying—it would be warm work "in front, they had had enough of it the day before * * *. "Presently—'Fall in 44th': cooking knocked on the head; "all was excitement. The brigade advanced under Pack, "and when we reached a rise in the road, we discovered the "French army covering the country, and the Prussians hotly "engaged with them at a distance, as we advanced down "the slope. The French were in line, with skirmishers in "the fields of rye, which was about five feet high. We "advanced with the light company extended, but finding "that the French had the advantage of seeing us, and were

"picking off many, Colonel Hamerton called them in, and 1815
"file firing commenced from each company, to clear the rye "as we advanced. After several movements, the 44th were "detached at double quick to a rising ground, where we "found the French cavalry had driven our artillerymen from "their guns, and had taken possession of but could not move "them, as the horses were gone; many of our artillerymen "were sheltered under the guns. We were in quarter "distance column, and soon put our men in charge of their "guns again. A German regiment then came up, and the "44th rejoined their brigade. Soon afterwards the division "was in line on the plain: the roar of artillery was awful. "The French cavalry repeatedly charged, and we formed "squares: on the third occasion I was wounded."

Captain Burney was then carried to the rear, wounded in the head and leg. The achievement so graphically described by Captain Siborne occurred shortly afterwards.

The Prussians had been attacked the same day (16th of June) at Ligny, and were forced to retreat to Wavre, which caused the Duke of Wellington to make a corresponding movement, to keep up his communication with them. In the course of the morning of the 17th of June, the troops were withdrawn from Quatre Bras, and ultimately occupied a position in front of Waterloo, at a place named Mont St. Jean, without having been further molested in their march by the enemy than by his following, with a large body of cavalry, the cavalry under the Earl of Uxbridge, which afforded his lordship an opportunity of charging them.

At the beginning of the memorable battle of Waterloo, which commenced about ten o'clock in the morning of the 18th of June, the second battalion of the 44th, with the rest of Sir Denis Pack's brigade, was placed in support of some Belgian troops on the left of the main road to Brussels, and throughout the day was exposed to the fire of the enemy's artillery and sharpshooters. The left regiment (the 44th) was stationed on a knoll, in rear of the right of Best's Hanoverian brigade; and the 92nd, 42nd, and the third battalion of the Royals, stood on the right of the 44th.

1815 Marshal Ney having been ordered to attack the farmhouse of *La Haye Sainte*, advanced in force, and the brigade of Belgians of Perpoucher's division, which formed the first line of infantry, gave way at the mere sight of the formidable French columns, before they were within half-musket shot. Arrived in front of the British, a murderous fire commenced, fearfully thinning the nearest British division, which began to give way. Picton then ordered Pack's brigade—consisting of the third battalion of the Royals, 42nd, the second battalion of the 44th, and the 92nd to advance.

Sir Archibald Alison, in his "History of Europe," speaking of Sir Denis Pack's brigade at this point of the battle, states, "that it advanced with a loud shout, and "poured in so close and well-directed a fire, that the French "columns broke and recoiled in disorder."

While leading his division in this gallant charge, the noble Sir Thomas Picton received his death wound. "He fell gloriously," said his illustrious Commander, "leading his "division to a charge with bayonets, by which one of the "most furious attacks made by the enemy was defeated."

Napoleon now directed his strongest efforts to crush the left and centre of the allied army, with the view to cut them off from the possibility of a junction with the Prussians, who were known to be approaching from that direction; clouds of cavalry and of the far-famed Imperial Guard, snpported by a numerous artillery, were hurled against the regiments of Pack's and other brigades in this portion of the field, who were alternately thrown into square to receive them, and deployed into line as they drew off, the artillery opening fire.

The assaults of the French troops were calculated to spread confusion through any army; but, paralysed by the fierce determination of their opponents, the attacks of Napoleon's legions relaxed, and the Prussians at length effected a junction on the left to co-operate.

It was towards evening that two Prussian officers rode past, inquiring for the Duke of Wellington, and Blucher's columns began to appear moving upon the enemy's right, by the road from Wavre. At this time the French made a last grand effort by a general attack throughout the whole line, and the moment they were repulsed, the allied troops advanced, drove them from every position, and forced them to seek safety in flight, leaving their artillery, and all the *matériel* of an army, on the field. 1815

The 44th and other regiments were directed to desist from the pursuit and return to their original ground, which they did, giving three British cheers to the Prussian army. In this manner was achieved the Battle of Waterloo—the importance of which may be best estimated by the lengthened peace which ensued—and the memory of this eventful victory must ever survive as a proud monument of national glory and of British valour.

A repeater watch was taken on the 18th at Waterloo, by Ensign Dunlevie, of the 44th. When the regiment had reformed line from square, a French cavalry officer found himself the sole representative of his squadron, and hemmed in between two lines of troops. Whereupon he threw off his helmet, disguised himself in his cloak, and, being splendidly mounted, charged the rear centre of the 44th (first line), making a grasp at the colours. The sergeants called out, "here is a staff officer, open out"; on this, Ensign Dunlevie—who held one of the colours (and which the French officer made a snap at as he rode through)—stabbed the horse in the stomach; the animal staggered and fell about twenty yards in front; Dunlevie and two soldiers hastened on, and the Frenchman was bayoneted, while disengaging himself, pistol in hand, from his saddle. His watch and gold chain fell into their hands, and were afterwards purchased by Lieut.-Colonel Burney for thirty napoleons, their original value; the latter fact being discovered in the following manner:—Ensign Dunlevie subsequently took the repeater to the maker, in the Palais Royal, who referring to his book, said it had only been bought two months since for the above sum by a captain of dragoons, whose family had dealt with him for forty years. Only half the purchase-money had been paid, and so the watch-maker placed it at once in the drawer as security for the balance. There being an order

1815 from the Duke of Wellington forbidding any disputes with the French, and Dunlevie not speaking the language fluently, asked the shopkeeper to compare the watch with his time. On regaining possession of the prize, he carried it off, coolly wishing the Frenchman "good morning."*

At Quatre Bras the 44th Regiment lost 17 officers killed and wounded, namely:—Lieutenant William Tomkins and Ensign Peter Cooke were killed. The wounded were—Lieut.-Colonel John M. Hamerton, slightly; Captains Adam Brugh, David Power, William Burney,† and Mildmay Fane, severely; Lieutenants Robert Russell, Robert Grier, and W. B. Strong, severely; Lieutenant John Campbell, slightly; Lieutenant William Marcus Herne, severely; Lieutenant James Burke, slightly; Ensigns James Christie, and Benjamin Whitney, dangerously; James Carnegie Webster, and Alexander Wilson severely. Ensign Whitney, as at Bergen-op-Zoom, was again wounded in carrying the colour, and also received a severe bayonet wound at Waterloo.

* Colonel McMahon, in hunting up the particulars regarding the Salamanca Eagle, met with an old pensioner at Chelsea Hospital, named Thomas Brooks, who formerly served in the 44th, and lost an arm at Waterloo. He belonged to Captain Burney's company, No. 8, and was hit in five places while charging at Waterloo, at the close of the day. Brooks remembered the charge of the French officer at the colours, and the taking of the repeater above described.

† Captain Burney, who was severely wounded in the leg, and dangerously in the head, relates that, as he was being carried off the field by two sergeants, the Duke of Wellington and his staff passed by, when Captain Cole of the Horse Artillery, an old friend, stopped to take leave. Captain Burney having fainted, the other thought all was over, and gave directions for his being buried in the next ditch. A few years afterwards, Captain Cole, at a ball at Portsmouth, on seeing his friend enter the room, rushed out, and enquired if an officer of the name of Burney was one of the guests. Finding this to be the case, there was a happy reunion with the "dead alive."

Assistant-Surgeon Collins, after extracting the ball, put it into Captain Burney's pocket, having written on a piece of paper in pencil, that it had been taken from the head. The medical officers had all agreed that unless the bullet could be extracted, the sufferer must die mad. The ball, which was a little flattened, has been preserved as a precious relic.

On the 18th, Major George O'Malley (Brevet-Lieut.-Colonel), slightly; Lieutenant James Burke, Ensign Benjamin Whitney, and Adjutant Thomas M'Cann, severely. 1815

The casualties in non-commissioned officers and men, from the 16th to the 18th of June, amounted to one-third of the strength of the regiment, namely, fourteen killed, and one hundred and fifty-one wounded: on the 16th, one drummer and nine rank and file killed; and twelve sergeants and eighty-two rank and file wounded: on the 18th, four rank and file killed, three sergeants and fifty-four rank and file wounded. Ten of the wounded underwent amputation, and one hundred and twenty-five rejoined the regiment. The killed and died of wounds amounted to twenty-one non-commissioned officers and men.

Both Houses of Parliament most enthusiastically voted their thanks to the army "for its distinguished valour at Waterloo;" and the 44th, and other regiments engaged, were permitted to bear the word "Waterloo" on their colours and appointments.*

In acknowledgement of the services which the army performed in the battle of Waterloo, and the actions immediately preceding it, each subaltern, non-commissioned officer, and soldier present, was permitted to count two years' additional service, and silver medals were conferred on all ranks, bearing on one side an impression of His Royal Highness the Prince Regent, and on the reverse the figure of Victory, holding the palm in the right hand, and the olive branch in the left, with the word "WELLINGTON" over its head, and "WATERLOO, June 18, 1815," at its feet.

The following officers of the 44th were awarded the *Waterloo* medal:—

Lieut.-Colonel.—John Millet Hamerton.

Major.—George O'Malley (Bt. Lieut.-Colonel).

---

* On the 26th July, 1816, after the second battalion had been disbanded, the 44th was permitted to bear the word "Waterloo" on its regimental colour and appointments, in consequence of the indistguished conduct of the late second battalion in that battle.

1815

Captains.

John Jessop (Major).
Adam Brugh.
David Power.
William Burney
Mildmay Fane.

Lieutenants.

Robert Russell.
Ralph John Twinberrow.
Robert Grier.*
William Tomkins (killed 16 June).
William Burrough Strong.
John Campbell.†
Nich. Toler Kingsley.
James Burke.
Henry Martin.
William Marcus Hern.
Alexander Reddock.

Ensigns.

James Christie.
Benjamin Whitney.
Gillespie Dunlevie.
Peter Cooke (killed 16 June
James Carnegie Webster.
Thomas A. Sinclair.

Paymaster.—James Williams.
Adjutant.—Thomas McCann (Ensign)
Quarter-master.—Henry Jones.
Surgeon.—Oliver Halpit.
Assistant Surgeons. { John Collins
William Newton.

---

* There was a dog belonging to Lieutenant Grier, which was a great favourite with the regiment; it was pupped in the Peninsula, and may therefore be regarded as an old campaigner. The officer was a lieutenant in the light company, and the last time he was seen by Colonel Burney was in a room at Brussels wounded, with his faithful canine companion, also wounded, beside him.

† Lieutenant Campbell commanded the light company, and his father, Lieut.-General Archibald Campbell, had accompanied the Duke from Brussels to Waterloo, on a small grey pony, as a spectator. He remained in rear of the company with the feeling of backing his son in the performance of his duty, being anxious that he should distinguish himself. The balls flying about, and men falling rapidly, made the son equally anxious for the safety of his father; he implored him to go, as while he was present, it was impossible to perform his duty, or keep his eyes from him. The old general saw the necessity of complying with his son's wishes, and at once rejoined the Duke.

Major Fountain Elwin, Captains John Cleland Guthrie, and George Crozier, Lieutenants George Newberry and Robert Peacocke, were detained at Ostend as members of a general court-martial, the proceedings of which had to be approved at head-quarters before they could separate. Brevet Lieut.-Colonel Gregory was commandant at Ostend, Lieutenant Frederick Hemming was acting engineer there, and Lieutenant Temple Frederick Sinclair was town adjutant. All these officers, therefore, in consequence of being employed on other duties, were prevented from gaining the Waterloo medal. 1815

Lieut.-Colonel John Millet Hamerton, and Brevet Lieut.-Colonel George O'Malley were appointed Companions of the Most Honourable Order of the Bath, on the 22nd of June, 1815, and Captain Adam Brugh was promoted brevet-major, to bear date from the 18th of that month.

On the 19th of June the allied army resumed its triumphant pursuit of the French towards Paris, in the neighbourhood of which city the battalion encamped on the 3rd of July, without having been employed on any affair of consequence during the march.

The following General Order was issued by the Duke of Wellington on the 4th of July:—

"The Field Marshal has great satisfaction in announcing "to the troops under his command, that he has, in concert "with Field Marshal Prince Blucher, concluded a military "convention with the Commander-in-Chief of the French "army near Paris, by which the enemy is to evacuate "St. Denis, St. Ouen, Clichy, and Neuilly, this day at noon, "the heights of Montemarte to-morrow at noon, and Paris "next day.

"The Field Marshal congratulates the army upon this "result of their glorious victory. He desires that the troops "may employ the leisure of this day and to-morrow to clean "their arms, clothes, and appointments, as it is his intention "that they should pass him in review."

Louis XVIII. entered Paris on the 8th of July, and was

1815 once more reinstated on the throne of France. Napoleon Bonaparte having fled to the south of France, surrendered himself a prisoner, on the 15th of July, to Captain Maitland, commanding the "Bellerophon," British ship of war, and the island of St. Helena was afterwards appointed for his residence.

While in camp near Paris, the British army was reviewed by the Emperors of Austria and Russia, and other sovereigns in alliance with Great Britain.

Early in December the battalion marched from Montmartre, for Boulogne, where it arrived in the middle of December.

Prior to the removal of the battalion from Paris, the following General Order was issued :—

"*Head Quarters, Paris, 30th Nov., 1815.*

"Upon breaking up the army, which the Field Marshal "has had the honour of commanding, he begs leave again to "return thanks to the General Officers, and the officers and "troops, for their uniform good conduct.

"In the late short but memorable campaign, they have "given proofs to the world, that they possess, in an eminent "degree, all the good qualities of soldiers, and the Field-"Marshal is happy to be able to applaud their regular good "conduct in their camps and cantonments, not less than "when engaged with the enemy in the field.

"Whatever may be the future destination of those brave "troops of which the Field-Marshal now takes his leave, he "trusts that every individual will believe that he will ever "feel the deepest interest in their honour and welfare, and "will always be happy to promote either.

"(Signed) E. Barnes,

"*Adjutant-General.*"

1816 In the beginning of January, 1816, the battalion embarked at Calais for England, and by the 18th of that month the whole had arrived at Dover.

Prior to embarkation the news of the intended disbandment of the second battalion of the 44th had reached Calais, from which place the following memorial was addressed by the officers to His Royal Highness the Commander-in-Chief. 1816

"*To Field-Marshal His Royal Highness the Duke of* "*York, Commander-in-Chief, &c., &c., &c.*

"The memorial of the several officers now doing duty "with the second battalion of His Majesty's 44th Regiment "of Foot,

"Most humbly and respectfully showeth,

"That it is with feelings of the deepest regret they "observe by the public papers, that the second battalion of "their regiment has been ordered to be disbanded from the "25th ultimo, and as they observe that an arrangement has "been made sanctioning the continuance of several other "second battalions, at a reduced number of companies, your "memorialists, impressed with the highest sense of your "Royal Highness's wish and ardent desire to reward the "meritorious, are emboldened to call your attention to the "service of their corps.

"The second battalion of the 44th Regiment has been "about *twelve* years in existence, six of which it has been "actively employed in the Peninsula, France, Holland and "the Netherlands. It distinguished itself in the assault of "Badajoz, where it suffered a loss of one-half its numbers, "and where its colours were the first planted on the bastion "of that fortress. It was also engaged in the affair of "Fuentes d'Onor, as well as at Cadiz, and actively shared "the dangers and glory of the battle of Salamanca, where "the regiment captured one of the enemy's eagles. In the "retreat from Burgos it was in the fifth division, which "covered the retreat of the army, and on the 25th October "of that year, was engaged with the enemy from daylight in "the morning to twelve at night, defending and holding the "village of Muriel, which afforded the army an opportunity "of effecting its retreat; and in all those affairs, the conduct "of the second battalion 44th Regiment was such as to

1816 "induce the several generals commanding to express their "sense of its meritorious conduct in their general orders. "After this the regiment returned to England a skeleton, "and was again recruited under the command of the "late Honourable Lieut.-Colonel Carleton, and embarked "with that officer for Holland in the short space of *four* "months. It was, in a very hard winter, actively employed "at Merxem, and in the trenches before Antwerp, endeav-"ouring to effect the destruction of the enemy's fleet, and "soon after distinguished itself in the storming of Bergen-op-"Zoom, and occupied *nine* bastions of that fortress; the "regiment has, lately, been engaged in the ever-memorable "battles of Waterloo, where, for its number of men, its loss "was as great and as severe as that of any corps in the army.

"Your Memorialists having detailed the leading "features of the services of their second battalion, they "cannot, without severe and poignant grief, take leave "of the colours under which they have so successfully "fought and so severely bled (about *seventy* officers having "been killed or wounded), and they had fondly entertained and "cherished the hope to the last moment that the arrange-"ments of the public service might have saved the sacred "remains of their standards—the Ægis of their gallantry—"from oblivion. The confidence they have in your Royal "Highness's protection still keeps alive that hope; and "though the several individuals who come within the pale "of reduction—the generality of whom have borne the "fatigues, the dangers, and supported the glory of their "corps—feel satisfied that, as the public service admits of it, "they will be hereafter brought in from the half-pay list; "still they most *earnestly* and respectfully implore your "Royal Highness to continue on the establishment of the "army as many companies of the second battalion 44th "Regiment (if but two) as will keep it in existence, and as "will give protection to those colours under which, it is "hoped, the regiment may in future ages fight and "distinguish itself. Your Memorialists, in conclusion, beg "to say, they are the more forcibly induced to submit the "foregoing, and to solicit this request of your Royal

"Highness, from the positive circumstance that the second "battalion of the 44th Regiment is the only second battalion "of the army which served at Waterloo, and which is "named for total reduction. 1816

"The foregoing is most respectfully submitted.

"*Calais, 4th January, 1816.*"

This memorial was signed by the officers doing duty with the second battalion, and the following is the reply received from the Military Secretary :—

Horse Guards, 22nd January, 1816.

"Sir, I have received and laid before the Com-"mander-in-Chief your letter of the 4th instant, and its "enclosed Memorial from the officers doing duty with "the second battalion 44th Regiment, praying, for the "reasons therein assigned, that the corps should not be "included in the reductions rendered necessary by the "termination of the war.

"I have His Royal Highness's commands to assure "you, and the officers whose Memorial you forward, "that he is strongly impressed by the merits and im-"portant services which have distinguished the second "battalion of the 44th Regiment, and he sincerely regrets "that an impartial consistency with the principle adopted "for regulating the reduction of second battalions must "preclude the possibility of any exemption being granted in "favour of that corps.

"Were it possible to execute with general satisfaction "to the army or advantage to the public service the "invidious principle of retaining corps on the establishment "according to an estimate of their merits and services, "the second battalion of the 44th Regiment would "certainly have every fair claim to favourable consideration ; "but as such a principle (besides other insuperable reasons) "would have been at variance with the views to efficiency "which have alone regulated the selection of second

1816 "battalions that are to be kept up for a time, to supply the "force required for the defence of the empire, that corps "necessarily and unavoidably came within the scale of "reduction, from its appearing that, after completing the "first battalion (three hundred and nine rank and file) it "would not, in fact, have any effective men upon the "establishment, to justify its being retained.

"I have, &c.,

"(Signed) H. TORRENS.

"*Lieut.-Colonel Hamerton,*

"*Commanding Second Battalion 44th Regiment.*"

The second battalion was accordingly disbanded* at Dover on the 24th of January, 1816, the officers receiving full-pay to the 24th of March following; the whole of the men fit for service—consisting of fifty-one sergeants, twenty-two drummers, and seven hundred and eighteen rank and file—were transferred to the first battalion.

* The officers of the second battalion, on its being disbanded, had the silver plate of their mess converted into a very handsome soup tureen, which they presented to their brother officers of the first battalion. This, and Captain Fyffe's snuff-mull before alluded to, were fortunately left behind by the corps when it marched to Cabul in 1840, and are still in the officers' mess.

SERVICES OF THE REGIMENT.

Continued from PAGE 53.

During 1816 and the five following years, the regiment 1816
continued in Ireland.

On the 28th of January, 1820, Lieut.-General Gore 1820
Browne was appointed Colonel of the 44th, in succession to General John, Earl of Suffolk, deceased. New colours were shortly afterward presented to the regiment.

In January, 1822, the 44th proceeded from Athlone to 1822
Dublin, and occupied the Royal Barracks. Early in April, the regiment, having been placed under orders for India, was moved to Chatham, and, on the 7th and 8th of June following, embarked at Gravesend, under the command of Colonel Morrison, in the Honorable East India Company's ships "Winchelsea," "Warren Hastings," and "Dorsetshire," for Calcutta. The strength of the regiment consisted of thirty-seven officers and six hundred and thirty-eight non-commissioned officers and men. It was disembarked at Calcutta between the 1st and 20th of November, and marched to Fort William; here one hundred and ninety volunteers, from corps about to return to England, considerably raised the strength of the regiment, and on the 25th of December it mustered eight hundred and nineteen non-commissioned officers and men, nine casualties having occurred since its arrival.

On the first of July, 1823, the headquarter division, 1823
consisting of five companies, proceeded from Fort William to Dinapore, where it arrived on the 16th, 17th, and 18th of August. Eight men were unfortunately drowned by the upsetting and sinking of nine boats in a storm, while on a passage up the Ganges. On another occasion, on this passage, a fire broke out in the leading

1823 boat of the fleet, whilst moored for the night at the bank of the river. The wind blowing down stream, the fire spread rapidly along the line of boats, and besides doing other damage destroyed the whole of the band instruments, and music of twenty years' collecting. The Regiment was, in consequence of this accident, deprived of its band for a year.

Already the regiment had suffered severely from the climate af India; at one time seventeen officers were laid up with fever, and Major John Cleland Guthrie, Captain Philip O'Rielly, and Lieutenants Ralph John Twinberrow and William Sargent, died at Calcutta, between November 1822 and June 1823.

The left wing arrived at Dinapore on the 22nd of September, and landed on the following day, having quitted Fort William on the 8th of August. During the remainder of the year the 44th continued at Dinapore, under the command of Colonel Morrison, C.B.*

1824 The right wing embarked in boats at Dinapore on the 31st of March, arrived at Bogwangola on the 12th of April, and reached Berhampore on the 21st and 22nd of that month; the left wing, under Major Carter, proceeded by water to Fort William. On the first of June the head-quarters were removed to Fort William, the five companies which had been stationed there having been ordered to Chittagong, where, in consequence of war having been declared against the court of Ava, a force was being assembled. Colonel Morrison, of the 44th, was appointed to command this force, with the rank of Brigadier-General. Another officer of the regiment, Colonel John Henry Dunkin, C.B., was also appointed to the command of a brigade on the

---

* It may not be altogether without interest here to record that there was at this time serving in the regiment an Ensign Edward Gilbert, who had been promoted from the ranks for service during the late war. He died in India in 1824, leaving a widow and an infant daughter: the latter in course of time married an officer of the Indian Army, but was separated from her husband, and became afterwards celebrated under the assumed name of "Lola Montez."

Sylhet frontier. The right wing, under Captain William 1824
Burney,* embarked at Fort William for Chittagong on the 26th of November, and the regiment was altogether on the 17th of December.

The force which had been assembled at Chittagong, under Brigadier-General Morrison, consisted of eleven thousand men, and comprised the 44th and 54th Regiments of the line; the 26th, 42nd, 49th, and 62nd, Bengal Native Infantry; 2nd Light Infantry battalion; the 10th and 16th Madras Native Infantry; the Mug levy, and a body of local horse, with artillery and pioneers. The object in view was to take possession of the province of Arracan, and to reach the upper part of the Irrawaddy, in the vicinity of the capital, by a route across the chain of hills forming the western boundary of that river.

Meanwhile the army under Brigadier-General Sir Archibald Campbell, K.C.B., had proceeded to carry out the design of occupying Rangoon and the country at the mouth of the Irrawaddy. By the end of the year 1824 several actions had taken place with the Burmese, and some important provinces had been captured.

At this period of the war the force under Brigadier- 1825
General Morrison moved forward to effect a junction with the

---

* Afterwards Colonel Burney, K.H.; he served for many years with the regiment, and much valuable information was obtained from him for this record, in which his services appear in the order of date. During this Burmese campaign he was present at the attack of the Padawa Pass, the fortified positions of Mahattee, storming the stockades and hills near Arracan, and capture of that city and its works. He led the advance in the latter operations on the 1st April, 1825. After quitting India in a most impaired state of health, from repeated attacks of fever, and the climate acting on the wound in his head received at Quatre Bras, he was removed in 1821 to the 75th Regiment. In 1834 he was appointed to the command of the Cape Mounted Riflemen, and was employed as second in command of the cavalry division, during the war against the Caffres. He received the war medal with clasp for Fuentes d'Onor, and the Waterloo, Burmese, and Cape medals.

1825 army at Rangoon. The troops left Chittagong early in January, 1825; some regiments proceeded by sea, others in boats along the coast, but the 44th, under the command of Major Carter, went by land, and on the 25th of that month was encamped at Ramoo, the stockades of which place were stormed. An attack was also made on Ramree on the 3rd of February. The enemy was driven at the point of the bayonet from all his intrenched positions, and was compelled to take shelter in his usual place of resort, the jungle, from which he kept up a galling fire until dislodged by the advancing troops. It being found that the jungle was so dense as to be impenetrable, and that it was moreover lined by the enemy, who had opened a fire upon the rear, orders were given to return to the boats, and the force was re-embarked by six o'clock in the evening. No loss was sustained on this service by the 44th Regiment.

In consequence of the inclemency of the weather the troops were compelled to put back to Mungdoo, but on the the evening of the 23rd of February they again sailed in transports (having previously been embarked in gunboats) for Arracan River, where they anchored in the evening of the 26th of that month. On the 5th of March they were disembarked, and commenced their advance upon Arracan. In about three weeks the 44th arrived at Chabatta, within two miles of the enemy's position in the Padawa Hills. On the 25th of March boat and pontoon bridges were thrown across the Chabatta and Waibrang rivers, and in the evening a reconnaissance was made to ascertain the situation of the passes through the mountains, and the obstacles to be encountered. The natural difficulties were considerable, owing chiefly to the steepness of the ascents, and to the courses of deep tidal nullahs. On the 26th, at daybreak, operations were commenced to force the passes. After the advance had gained the summit of the hills, which were occupied by the Burmese in great force, the centre column moved to its left to attack a stockade, whilst the light infantry companies, keeping almost parallel with it, drove the enemy from several intrenched positions along the crest of the hills; two rounds from the twelve-pounders

caused the abandonment of the partly-finished stockade, but 1825
the men who retired from it immediately occupied strong ground on the heights above; here the left column, from being compelled to diverge from the river, joined, and a smart fire was continued, until the grenadiers of the 44th Regiment and of the 49th Bengal Native Infantry began to ascend at two different points; the enemy soon sought refuge in flight, and, taking advantage of the nature of the country, for a time escaped.

The forward movement was continued, and the heads of the three columns, which had become united at the Jeejah river, no sooner debouched from the jungle than the enemy fled, and the troops reached, with little loss, the works that covered the fords of the Mahattee. Having thus forced the passes of the Padawa Hills, the ford was crossed on the 27th of March. As soon as a fog which concealed the enemy from view had dispersed, Major Carter, with three companies of the 44th, covered by the light company of the 54th, was directed to carry a small hill in front of the works, whilst the remainder of the force moved onward in quarter distance column, with the artillery on the reverse flank. The enemy's position was well chosen, being situated on a peninsula, protected by a broad river, whose fords were only passable at nearly low water, the banks being steep, and thickly planted with sharp stakes. The defences consisted of deep intrenchments along the margin, with epaulments to protect them from enfilading fire; in the rear were high conical hills, surmounted by pagodas, and surrounded by intrenchments, like so many citadels, and appearing to be so occupied.

Upon Major Carter's advance, a large intended reinforcement of the enemy was seen approaching, but it did not remain long in sight, nor reach Mahatee. All opposition was very soon overcome, but it became necessary to halt on the 28th, in order to obtain information, and to enable the troops in the rear to come up. This was effected in the evening, when a partial reconnaissance of the position was made.

The approach to Arracan was across a narrow valley,

1825 bounded by a range of hills about four hundred feet in height, the summit being defended by a series of stockades and a garrison estimated at nine thousand Burmese.

An unsuccessful attack was made upon the place on the 29th of March, and on the evening of the 31st Brigadier Richards, with a detachment, consisting of six companies of the 44th, three of the 26th, and three of the 49th Bengal Native Infantry, thirty seamen, and a like number of Gardner's dismounted horse, ascended a lofty range by a circuitous route, and succeeded in establishing himself on the summit, before the enemy detected the movement. On the following morning the Burmese were attacked in flank by this division, whilst they were assailed in front by the main body. Driven from each stockade, the enemy adandoned Arracan, and withdrew over the country between the city and the mountains, passing the latter by the Talak and Aeng defiles. Thus, with trifling loss, was the first object gained,—the casualties in the 44th, from the 26th of March to the 1st April, being limited to one sergeant and twelve men wounded.

On the 7th of April, four companies of the regiment, commanded by Captain John Shelton,* and consisting of Captain John Connor, Lieutenants Benjamin Whitney,†

---

* This officer, afterwards so intimately connected with the 44th, was promoted to the rank of Major, on the 28th of April, 1825,—Major Adam Brugh having died in February, whilst on passage to England.

† Lieutenant Whitney was promoted to a company in the 44th on the 28th of April, 1825. He served as a volunteer with the 43rd in the Peninsula, and was severely wounded when crossing the Mondego. On the 25th of February, 1813, he was appointed ensign in the 44th; and served the campaign in Holland in 1814, with the second battalion; was present at both actions at Merxem, and highly distinguished himself at Bergen-op-Zoom, and at Waterloo, as detailed in the foregoing pages. During the first Burmese war, he was present at the storming of the stockades at Ramoo, Ramree, the intrenched positions on the Padawa Mountains, and at Mahattee. He also led the storming party at the stockades on the heights of Arracan, and was present at the siege and capture of that city on the 29th, 30th, and 31st March, and 1st April, 1825. This officer

John Donaldson, and Thomas Robinson, Ensign Robert 1825
Bradford M'Crea, and Assistant-Surgeon Richard Verling, M.D., twelve sergeants, four drummers, and one hundred and ninety-four rank and file, were sent to Ramree, an island about one hundred miles lower down the coast,—the country being of the same mountainous and jungly character as that already travelled over. From this expedition the troops returned about the middle of May, after having been at sea, or on shipboard, for nearly six weeks, during which period only two men were lost, and those from disease. It was otherwise, however, with a company of the regiment consisting of Captain Thomas Mackrell, Lieutenant James Paton, two sergeants, one drummer, and fifty-four rank and file, which proceeded on the 12th of May, to take possession of Talak, a small post situated at the foot of the hills, about thirty miles to the south-east, and to which the detachment proceeded, partly by inland navigation and partly by most fatiguing marches through a dense and swampy jungle. After suffering severely from fever, the detachment was on the 7th of June withdrawn, but not before several men had died; many more being besides lost to the service from the same malady, within a week or two after their return to head-quarters.

Upon the monsoon setting in, fever began to augment, both in prevalence and severity, and by the end of October, one hundred and forty-seven men had died. Some idea of the state to which the regiment was reduced may be formed from the circumstances that, of the four hundred and twenty-seven men mustered with the 44th on the 25th of November, two hundred and three of whom were on the sick list, one hundred and seventeen died before the same day of the suc-

was promoted major in the 14th foot (to which he had exchanged) on the 25th of November, 1838, and, in 1838, whilst serving in the West Indies, commanded the garrisons of Brimstone Hill in the Leeward Islands, and in the following year that of Dominica. He was presented with a sword by the Revenue Board in Dublin, and received a letter of thanks from the Magistrates of the counties of Dublin, Kildare, and Wicklow. Major Whitney retired in 1840, and died in 1862. He had received the Waterloo medal and that for the Burmese campaign of 1825.

1825 ceeding month, and two hundred and fifty-three of the survivors were in such a condition, that they had to be sent, upon landing, to the General Hospital at Calcutta.*

When the fever first prevailed, the maximum range of the thermometer was from 90° to 93°, with light air and cloudless sky, a degree of temperature which, in a locality surrounded by hills, was very oppressive; but the disease became even more prevalent and fatal when the weather got cooler.

Under these sad circumstances, it became necessary to withdraw the troops, to prevent them all falling a sacrifice to the pestilential climate; the regiment, therefore, commanded by Major Carter, left Arracan on the 1st of December, and arrived at Fort William on the 16th of that month. Whilst on the voyage, about twenty cases of cholera broke out, most of which proved fatal.

Lieut.-Colonel and Colonel John Henry Dunkin, C.B., who was serving as brigadier on the Sylhet frontier, died at Dacca, on the 11th of November, and was succeeded in the Lieut.-Colonelcy by Major Carter. Lieutenants John Donaldson, Albert Grant Gledstanes (adjutant), Henry Dick Carr and James Paton, and Ensign Hemsworth Ussher, together with Brigadier-General Morrison, who had been compelled to leave in October, on account of ill-health, fell victims to the climate of Arracan.

The 54th suffered in like manner with the 44th, and the appearance of the two regiments on their arrival at their respective Presidencies exhibited a melancholy instance of the baneful effect of the climate upon European constitutions. The number of deaths and sick exceeded even that of the

* During the first four months of 1825, twenty-nine deaths occurred, but during the remainder of the year, until December, two hundred and sixty-four men died, namely:—

In June	24	In September	17
In July	30	In October	42
In August	34	In November	117

264

Walcheren expedition, in proportion to the strength of the force. Nearly every soldier had to be immediately placed under hospital treatment, his chance of recovery resting mainly upon a speedy return to England; numbers, however, died before that remedy was available. Of those who left Arracan, scarcely one half were alive at the end of twelve months.* 1825

Whilst the regiment was on its return to Calcutta, the Burmese forces had been driven from their strongly fortified positions; this induced them to sue for peace; but it being afterwards discovered that this was only a device to gain time to reorganise their troops for a more determined resistance, hostilities were in the middle of January, 1826, resumed, and on the Anglo-Indian forces appearing before Ummerapoora, after having again defeated the enemy, the King of Ava sent the ratified treaty, ceding a considerable portion of territory, and agreeing to defray the expenses of the war. 1826

In the General Orders issued on the 11th of April, 1826, upon the termination of the Burmese War, honourable mention was made of the troops under Brigadier-General Morrison; after specifying the service of the army under Major-General Sir Archibald Campbell, it was added:—

"The Governor-General in Council cannot conclude "these General Orders, expressive of his high approbation "of the merits and services of the army under Major-"General Sir Archibald Campbell, without intimating, at "the same time, his entire satisfaction of the conduct of the "two divisions of British troops intended to penetrate into "Ava from our north-eastern and south-eastern frontiers, "and also of the British force employed in the expulsion of "the enemy from the country of Assam.

* A peculiar feature in the malady was the vomiting of disgusting worms, attributable to the worse than inferior quality of the flour, and provisions generally, issued as rations on the campaign. This was fully established by the report of a medical board ordered at Calcutta, to inquire into the cause of the excessive mortality amongst the troops employed.

1826 "The latter service—viz., the conquest of Assam—was "achieved by the force under Lieut.-Colonel A. Richards "with the most complete success; the capital, Rungpore, "having surrendered on terms, and the Burmese troops "having been entirely expelled from that country.

"On the side of Cachar, physical difficulties of an "insurmountable nature having arrested, at its very outset, "the progress of the army under Brigadier-General "Shuldham, no opportunity was afforded to that army of "displaying those qualities of courage, perseverance, and "zeal which the Governor-General in Council is satisfied it "possessed in common with its more fortunate brethren in "Ava.

"Similar and no less impediments ultimately opposed "the advance of the fine army under Brigadier-General "Morrison, over the mountains of Arracan, into the valley "of the Irrawaddy; but the capture, by the detachment "under Brigadier W. Richards, of the fort and heights of "the capital of Arracan, afforded an earnest of what would "have been effected, had opportunities offered, by the judg-"ment, prudence, and skill of the commanders and officers "of that division, and by the valour, zeal, and intrepidity of "the troops of which it was composed. The Governor-"General in Council deeply laments the general sickness "which attacked, and utterly disabled for further effective "service, the south-eastern division of the army, and the loss "of many brave officers and men, who fell victims to the "noxious climate of Arracan.

"In testimony of the high sense entertained by Govern-"ment of the services of the troops by whom the provinces "of Assam and Arracan were conquered, the Governor-"General in Council is pleased to order that the several "Native corps who were employed in those countries shall "respectively bear on their colours the words 'Assam' and "'Arracan,' as the case may be, and His Majesty will be "solicited to grant to the 44th and 54th Regiments the "same distinction."

"Assam" and "Arracan" were not authorised for the

King's troops, and the 44th and other Regiments of the line 1826
had the word "Ava," in commemoration of the foregoing
services, on their colours. When the retrospective war-
medal for Indian services was authorised, a clasp was granted
for "Ava."

On the 14th of January, 1826, the regiment proceeded from Fort William to Ghazeepore, where it arrived on the 4th of April, and remained upwards of two years.

In October, 1828, the 44th left Ghazeepore for Cawn- 1828
pore, which it reached towards the end of November.

During the following five years, the head-quarters 1829
remained at Cawnpore, but in December, 1833, the regiment to
marched for Chinsurah. 1833

The regiment arrived at Chinsurah in February, 1834, 1834
and continued there until January, 1835, when it proceeded 1835
to Fort William.

During the year 1836, the regiment remained at Fort 1836
William, but proceeded in January, 1837, to Ghazeepore, 1837
and arrived at the station in the following month. It
suffered much from cholera while quartered there, two
hundred and fifty men dying within three months.

The regiment remained at Ghazeepore until November, 1838
1838, when orders were received to march to Kurnaul,
where it arrived in January, 1839. 1839

Shah Soojah-ool-Moolk having in August, 1839, been
reinstated on the throne of Affghanistan, orders were issued
in January, of the following year, for breaking up the 1840
"Army of the Indus," and for the return of the troops to
their several stations. A force had, however, to be left in
Affghanistan, and some fresh regiments were ordered to proceed from India as reinforcements and reliefs.

Amongst those selected was the 44th Regiment, which was formed in brigade with the 5th and 54th Native Infantry, and numbered the third infantry brigade, Lieut.-Colonel John Shelton being placed in command of these corps, with the rank of Brigadier-General.

Thus was the regiment and its commanding officer

1840 brought into connection with a land full of ancient memories, and wherein it was destined very soon to find its grave; before this came to pass, however, opportunities were afforded the 44th of gaining an undying name, and although the sequel was unfortunate, the corps passed through an ordeal—trying as it undoubtedly was—in a manner which has added to the renown of the British Army.

In October, 1840, the regiment quitted Kurnaul, reached Char-Deh, en route to Cabul, in December, and in
1841 January, 1841, arrived at Jellalabad, a place soon to be rendered immortal in military history.

The restoration of Shah-Soojah to the throne being a most unpopular measure with the six Sirdars of Cabul, the country was in a very disturbed state. No sooner had Brigadier Shelton arrived at Jellalabad, than he had to proceed to the Nazian Valley, to coerce a refractory tribe named the Sungho Khail. The force selected consisted of the 44th, under Lieut.-Colonel Mackrell, the 27th Native Infantry, a troop of horse artillery, a detachment of sappers and miners, and a portion of Shah-Soojah's cavalry and infantry. It marched from Jellalabad on the 21st of February, and three days afterwards nearly the whole of the forts (upwards of eighty in number), which studded the Nazian valley, eight miles in length, were captured. The valley is described as being particulary rugged, and frequently narrowing in with perpendicular and almost inaccessible heights. The services of the force were acknowledged in Government Orders, as having been performed with a degree of perseverance, daring, and exertion, highly creditable to the Brigadier, and the officers and men under his command.*

Having effected the object of the expedition, Brigadier Shelton returned to Jellalabad on the 17th of March. On the 8th of May, the Brigadier having received intelligence that the family of the King, on its way to Affghanistan from India, was intercepted at the Indus by some refractory Sikh battalions, moved his brigade rapidly back through the

---

* Brigadier Shelton received from Shah-Soojah for this service the second class of the Douranee Order.

Khyber pass, surprised and dispersed them, and left the road open for the King's family to pass unmolested. The Brigadier then retraced his steps, and marched on to Cabul, where he arrived on the 9th of June, and was encamped at Seah Sung (Black Rock), near that city. 1841

The 44th remained in this encampment until October, without any occurrence of note, except that four companies of the regiment, under Major Scott, were sent in September with a force under Lieut.-Colonel Oliver, 5th Native Infantry, to the Zoormut Valley, to coerce another refractory tribe. This object having been accomplished, the force returned to cantonments about the 14th of October. A crisis was however approaching, and the government of the restored monarch became so unpopular, that the Affghans resolved to effect the expulsion of the British, whose presence was naturally unwelcome to them. Major-General Sir Willoughby Cotton had been succeeded in the command of the forces in Affghanistan in April of this year by Major-General Elphinstone; but this latter officer, although a most estimable man, was, unfortunately, from age and ill-health, unequal to the emergency, which required great decision of character and promptness of action.

On the 1st of October the strength of the regiment at Cabul consisted of twenty-five officers,* thirty-five sergeants,

---

* The officers of the 44th, at Cabul, were—

Lieut.-Colonel Shelton, commanding third infantry brigade, as brigadier.
Lieut.-Colonel Mackrell.
Major Scott.
Captain Swayne.
,, M'Crea.
,, Leighton.
,, Robinson (brigade-major to third infantry brigade).
Lieutenant Dodgin.
,, Collins (promoted to an unattached company, on July 16th, 1841).
Lieutenant Evans.
Lieutenant White.
,, Souter.
,, Wade.
,, Hogg.
,, Cumberland.
,, Raban.
,, Cadett.
,, Swinton.
,, Fortye.
Ensign Gray.
Paymaster Bourke.
Quarter-master Halahan.
Surgeon Harcourt.
Assist.-surgeon Balfour.
,, Primrose, M.D.

The above, with the exception of Colonel Shelton, Lieutenants

1841 fourteen drummers, and six hundred and thirty-five rank and file. It is important to notice this, since it will be seen that all the officers save three were killed, and nearly the whole of the men perished.

Early in this month (October) the Khoord Cabul Pass was blocked up by certain refractory chiefs, their hostility being imputed to the fact that the allowances made to them had been reduced. Troops were detached on the 9th of October, under Lieut.-Colonel Monteath, C.B., and two days afterwards Major-General Sir Robert Sale, K.C.B., Lieut.-Colonel of the 13th Light Infantry, proceeded with a small force, of which that regiment formed a portion, with the object of clearing the Pass, and re-opening the communication with India. The Major-General, upon receiving intelligence of the events about to be related, made two forced marches upon Jellalabad, it being impossible to return to Cabul, and gallantly defended the dilapidated fortress until relieved by Major-General Pollock.

On the departure of the 13th Light Infantry the 44th was moved from camp at Seah Sung into cantonments.

At this period it is futile to refer to what might have been; nevertheless it is certain that, although it was hoped that it would blow over, there were sure indications of the coming storm. This imaginary security was dispelled on the 2nd of November, when the threatened outbreak burst forth at Cabul, and the political agent, Sir Alexander Burnes, residing in the city, was cruelly murdered, his house sur-

---

Evans and Souter, were all killed; Lieutenant Souter being the only officer who was present at and survived the final struggle.

Captain A. W. Gray and Lieutenant R. Kipling (Adjutant) had obtained leave to Europe, and left Cabul for India on the 2nd of October. *En route* through the passes of Affghanistan they were molested and fired upon, and were obliged to adopt the Affghan costume for safety. Lieutenant Kipling, who had been Sergeant-major, and, though a fine old soldier, was no linguist, would most certainly have been murdered, but for his companion, Captain Gray, whose appearance and proficiency in the language, well enabled him to pass for a native chief; the two, after many narrow escapes, at length reached the plains in safety.

rounded and set fire to, and he himself, with his brother and 1841
Lieutenant Broadfoot, consumed in the flames. The garrison in cantonments on this day consisted of Her Majesty's 44th, 5th Native Infantry, and six companies of the 54th Native Infantry. Brigadier Shelton, who was encamped at Seah Sung, with three companies of the 54th Native Infantry, two squadrons of the 5th Light Cavalry, together with Captain Nicholl's troop of horse artillery, the 6th Regiment of the Shah's light infantry (recently embodied), and the 2nd Regiment of the Shah's cavalry, received orders to proceed with a detachment, in which was included one hundred men of the 44th under Lieutenant Souter, to the Balla Hissar (high fort), the citadel of Cabul, into which he accordingly marched. This was a fortress of imposing appearance, and, when first built, must have been a place of great strength; but its walls were fast crumbling to decay. The remainder of the troops were concentrated in cantonments, from whence a small force was sent on the following day to co-operate with Brigadier Shelton by an attack on the Lahore gate of the city. This was opposed by so large a body of the enemy that it failed in reaching its destination. Three hundred of the Shah's own soldiers, with two guns, were also sent into the city to suppress the insurrection, but they, likewise, did not succeed in their object.

The stores of grain laid in for the Shah's own force, being close to the city, fell at once into the possession of the insurgents. The magazine fort, containing all the supplies for the British force, and the European hospital stores of wine, beer, medicines, &c., and twenty days' issue of grain—all that had been laid in—was only a few hundred yards from cantonments, and occupied by a guard of Sepoys, under an European officer who reported on the 4th of November, that unless reinforced, he should be compelled to retire. The place was commanded by the enemy from the old fort at Mahomed Shereef and from the Shah's garden immediately on the opposite side of the road, rendering communication by the usual entrance exceedingly difficult. Captain Thomas Swayne, of the 44th, was sent out with his single company to endeavour to effect the required relief,

1841 and fell gallantly at the head of his men, many of whom were shot down at the wall, and the remainder at last driven back, the officer in command of the fort having so effectually barricaded the gate of his fort with camel saddles three deep that he could not give admission to the reinforcement he had called for.

Captain Thomas Robinson, also of the 44th, who had been serving as brigade-major to the third infantry brigade since September, 1841, then proceeded with his company for the same purpose, and with a like result, for he also lost his life. Lieutenant Arthur Hogg and Quartermaster Halahan were wounded on this occasion.

Early on the following morning, the 5th of November, another effort was resolved upon, but by the time the troops received orders to be in readiness, it was found that the officer and his guard had abandoned the place, having retired into cantonments by withdrawing through a hole that had been made in the wall, and through which the Sepoys had one by one gradually left him. In this manner the fort fell into the enemy's hands, together with the only remaining store of provisions.

It was now determined to attack the fort at Mahomed Shereef, which, if it had been taken on the previous day, would have saved the Commissariat Fort. The first attempt failed, but on the following day (the 6th) it was again attacked and this time captured. Lieutenant G. Raban, of the 44th, was killed on the top of the breach, whilst gallantly leading his little party of the regiment.

Meanwhile, the enemy had become emboldened and encouraged, so that when Brigadier Shelton went into cantonments, on the 9th of November, it was too late to make any attempt upon the city, then filled with armed men, amounting to about thirty thousand; the British, moreover, had only three days' provisions.

The Affghans took possession of the Rika-bashee and other forts close to the cantonments, early on the morning of the 10th of November, and made lodgments so near to the works in that quarter, that it was unsafe for the soldiers

to show their heads above the parapet. It was therefore 1841
determined, on the afternoon of that day (10th), to storm the Rika-bashee fort, when several men effected an entrance; the Affghans, meanwhile having assembled in considerable force, the retreat was sounded, and on the troops being withdrawn, the enemy's cavalry charging round the corner of the fort, killed many, amongst whom were Lieut.-Colonel Thomas Mackrell* and Captain Robert Bradford M'Crea, of the 44th. The former received several bullet wounds, besides sabre cuts, but survived until he reached his quarters, where he died, after having his arm amputated; one of his legs was also broken; Captain M'Crea was cut to pieces. The assault was renewed, and the place ultimately captured, together with two small adjoining forts, but not without a loss of two hundred men killed and wounded out of six hundred. This fact of every third man being struck evinces the skill of the Affghans, who were armed with excellent rifles. Their casualties were heavy, but it was difficult to ascertain the extent, from their great exertions in carrying off their killed and wounded. As a number of their dead were, nevertheless, left on the field, their loss must have been very considerable. These forts yielded a few days' provisions and forage, and their capture obliged some of the people in their neighbourhood to supply more.

On the 13th of November the Affghans opened two guns from the heights to the north of the cantonments; and apprehension being therefore entertained of a more formidable gathering of them, which might endanger the forts now in possession of the British, Brigadier Shelton sallied forth with a party, spiked one gun and brought the other into cantonments.

* Lieut.-Colonel Mackrell had been born in the regiment, and with him perished the last link, as it were, of the 44th with the memorable campaign in Egypt. It was told of him that he, then a drummer-boy, and completely knocked up with marching over the heavy sands, was carried into action on the back of his father, the Drum Major. He was given a commission in the regiment on the 19th September, 1804, and had slowly indeed, but at length, risen to its highest grade. A reminiscence, also, of Egypt, a mess-table ornament, a trophy of the campaign, was lost, with the whole of the mess property, on the disastrous retreat.

1841 Major Pottinger and Lieutenant Houghton, adjutant of the Gorkha regiment stationed at Charekar, in Kohistan, whose corps had been cut up, came into cantonments on the 15th of November, both wounded. A small affair took place on the 22nd of November, near the village of Be-maroo,* which was enclosed by a strong lofty wall, and it was resolved to make a more decided attack on the following day.

For this attack on the heights of Be-maroo there were told off about one thousand three hundred and fifty men; of these only nine hundred were infantry, and with but one gun. Brigadier Shelton, who commanded, applied for reinforcements and another gun, but they were not sent.

At two o'clock, a.m., on the 23rd of November, the force which had been selected for the service moved out of cantonments. It consisted of one horse-artillery gun, under Sergeant Mulhal; five companies of the 44th, under Captain Thomas Richard Leighton; six companies of the 5th Native Infantry, under Lieut.-Colonel Oliver, and a like number of the 37th Native Infantry, under Major Kershaw, of Her Majesty's 13th Light Infantry; a squadron of the 5th Light Cavalry, under Captain Bott, and one of the Irregular Horse, under Lieutenant Walker; one hundred of Anderson's Horse, and a like number of Sappers, under Lieutenant Laing.

Proceeding in perfect silence by the Kohistan Gate, the troops marched to the gorge at the further extremity of the Be-maroo heights, and notwithstanding the difficulty of the ascent, dragged the gun to the top. At daybreak, after an unsuccessful attempt to take the village, which was enclosed within high walls, parties were seen making off from the village to a distant fort; large bodies of men were also perceived advancing from Cabul, and eventually the British position was surrounded at all points, except that facing the cantonments. By this time, about ten o'clock in the morn-

* Be-maroo signifies husbandless, or without a husband; and this hill is so named from its being the spot where the vestal virgins were buried.

ing, the ammunition was nearly expended and the soldiers, 1841
wearied and faint from want of water, sustained this hard struggle until after twelve, without the desired reinforcements and additional gun being sent up. At length the troops had to retire, after suffering great loss. Lieutenants William Evans and Samuel Swinton, of the 44th, were on this day both wounded. No further attack was made by the British and although the Affghans frequently crowned the Be-maroo heights, they generally retired towards evening.

The fort of Mahomed Shereef, garrisoned by a subaltern's party of the 44th, and a company of Native Infantry, was unfortunately surprised on the morning of the 6th of December. The officer, Ensign Gray (44th), having committed the error of himself leaving the point of attack, and hastening to the side nearest to the cantonments, which latter were but just across the road, to call for a reinforcement, the garrison was left without an officer to enforce order, and, being awed by the number of Affghans coming on to the attack, withdrew, and the fort was immediately taken possession of by the enemy. Ensign Gray, who was the last man to leave the fort, in the attempt to retrieve his mistake, was severely wounded. Private John Stuart, of the 44th, on sentry at the time, refused to abandon his post, throwing hand grenades, firing his ammunition, and using his bayonet against the advancing enemy, and was eventually cut to pieces at his post, in the honourable discharge of his duty. Each company of the regiment came forward and volunteered to retake the fort, but it was not deemed advisable to do so.

Negotiations were shortly afterwards opened with the insurgent chiefs, who agreed to furnish supplies, and allow provisions to be brought into cantonments, on condition that the country was evacuated. This was agreed to, and on the 13th of December the 54th Native Infantry and guns were withdrawn from the Balla Hissa. No sooner were they out than the gates were closed by the Shah's troops, to prevent a party of Akbar Khan's followers from rushing into the place.

In this awkward predicament a message was received

1841 from Akbar Khan, the representative of the six Sirdars, who was to superintend the safe conduct of the British, stating that they must remain where they were, as the Ghilzyes on the hills were more than he could control. During the whole of the night the 54th Native Infantry was kept in the cold, and the poor animals with their loads on their backs. On the following morning the party was put in motion, when the rear was partially attacked on crossing the hill of Seah Sung, and several men were wounded.

During the afternoon of the 16th of December, the magazine and forts in proximity to the cantonments were given up, the Affghans marching in and taking possession as the British quitted them. A week afterwards, the fatal interview took place between Sir William Macnaghten and Mahomed Akbar Khan, and other Affghan chiefs, at which the British envoy was foully murdered; whereupon Major Pottinger was requested by Major-General Elphinstone to undertake the duties of political agent and adviser. At this period, news was received from Peshawur of the march of strong reinforcements from India. This prospect of relief was however but distant, and at the council of war which was convened, it was determined to renew the negotiations, and to pay to the chiefs the sums promised by the late envoy; several officers were, on the 29th of December, given up to them as hostages, and the greater portion of the treasure and guns made over to them. The remainder of the sick were sent into the city on the following day, under medical charge of Doctors Campbell and Berwick; Lieutenant Evans, who was wounded in the head on the 23rd of November, being in command. Thus ended the year, and under any but agreeable auspices was the festive season of Christmas passed—the snow on the ground alone reminding the soldiers of the seasonable weather of their own distant land.

1842 Nor did brighter prospects open with the new year, for the chiefs failed in their promise to provide two thousand camels and four hundred yaboos; some few were indeed sent, but, when required, even they were not forthcoming. It was at first intended to march on the 5th of January, 1842, but the chiefs having stated that their escort would not be ready until the day after, the departure was postponed. During the last fortnight the ground had been covered with snow, the thermometer being as low as zero. The force on leaving Cabul consisted of about four thousand five hundred fighting men, and the camp followers, women and children amounted to about three times that number: the strength of the 44th was nineteen officers and four hundred and thirty-eight fighting men, in which numbers were included Brigadier Shelton, twelve boys, and thirty-four sick. 1842

At nine o'clock in the morning of the 6th of January, the advance, under Brigadier Anquetil, consisting of the 44th, with some Sappers and Miners, one squadron of Irregular Horse, and three mountain train guns, moved off. The main column was commanded by Brigadier Shelton, and the rear-guard by Colonel Chambers. It was after midnight before the rear-guard reached Bygram, only five miles from Cabul; here the camp, if it could be so called, a few tents only being pitched, presented a sad picture of desolation, the face of the country being one uninterrupted scene of snow. Even Brigadier Shelton had no tent, and for six days and six nights was exposed on a bed of snow, with no other shelter than the canopy of heaven; a fire was out of the question, there being no wood to make one with. If such were the privations of officers high in command, it is not difficult to imagine those of others. 6 Jan.

The road, from the very gates of the cantonments, was strewed with the dead bodies of camp followers—men, women, and children, who, at this early stage in the retreat, fell victims to the frost and snow;—such was the intensity of the cold.

Daylight the next morning presented a scene of wretchedness and desolation which no description can fully portray; the sight of the stiffened remains of the poor creatures that had perished during the night from the inclemency of the weather was sufficient to stifle all hope in the bosoms of the fatigued and almost worn out survivors. 1842 7 Jan.

As the troops did not again move off until nine o'clock

1842 on the morning of the 7th, the Affghans had time to collect,
7 Jan. and the three mountain train-guns fell into their possession in making a detour at a blocked up portion of the road soon after starting. In a short time the rear guard (44th Regiment) was warmly engaged, and Lieutenant William George White, who was acting as Adjutant, severely wounded; several men were also killed and wounded. The road from Cabul was covered with Affghans following close after the British, and from time to time some portions of the baggage were left behind to distract the attention of the enemy, and give breathing time for a fresh start, but after a short interval they were again in pursuit.

Nearly the whole of the Shah's 6th Light Infantry of Hindoostanees had already deserted, and numbers of the Sepoys left their ranks and became intermixed with the baggage and camp followers. Some idea may be formed of the quantity of baggage, since on nearing Bootkhak, only about four miles from the last halting ground, it extended upwards of three miles. Here the best possible position was taken up, the 44th being detached to drive the enemy from the hills, which service the regiment effected, maintaining its position till night, when it was called in. Thus another terrible night had to be passed, without shelter or fire, on ground covered with snow.

1842 The succeeding morning (the 8th) displayed a horrifying
8 Jan. scene, many having fallen victims to the cold during the night. As soon as day broke the enemy's shot came flying amongst the camp followers and servants employed loading the baggage, which created great confusion. Brigadier Shelton hastened out just in time to repulse some Affghans that were pushing forward to the very ground occupied by the troops. Observing large bodies of them advancing, he hurried to the rear, where his regiment was drawn up in line, and just as the Affghans were closing, he gave the order to charge, when the 44th making a simultaneous rush, drove them back in gallant style, bayonetting many. This had a salutary effect, and kept the foe at a more respectful distance.

Soon afterwards negotiations were once more opened

with Akbar Khan, who demanded Brigadier Shelton as a hostage: this had been asked prior to the Envoy's death, but the request had been relinquished. It was now arranged that Major Pottinger, Captain Lawrence, Military Secretary to the late Envoy, and Captain Mackenzie, should proceed to the Sirdar as hostages. Shortly after twelve o'clock Akbar Khan sent forward about ten of his Affghan Horse to act as guides and protectors against the Ghilzyes. The troops moved on, the 44th and the 54th Native Infantry forming the rear-guard. The advance proceeded quietly on without other molestation than an occasional shot, until it reached the upper end of the Khoord (Little) Cabul Pass,* when the firing began and soon became heavy from one end of the Pass to the other. The confusion, slaughter, and plundering of baggage that ensued defies description. Major William Boxell Scott and Captain Thomas Richard Leighton, of the 44th, were this day wounded. In many places the Pass was so extremely narrow that it soon became blocked up with baggage, the dead bodies of men, women and children, and animals. The rear-guard made its way through the mass as best it could; but the Sepoys were so benumbed with cold that they were scarcely able to return a shot; the Affghans, in many instances, ran up to them and snatched their firelocks from their hands without any resistance being offered. It was observed that the Ghilzyes, the Affghan foot soldiers, did not attempt to injure those who were Mussulmen, being content with seizing their arms and stripping them of their clothes. No sooner, however, did a European fall into their power than they chopped him up with their large knives like a log of wood.

1842 8 Jan.

On clearing the Pass the Irregular Cavalry and Sepoys were so intermixed with the baggage and camp followers, that Brigadier Shelton could not collect two of them

---

* Amongst writers who should have been better informed, it has been stated that the massacre occurred in the Khyber Pass, which, being on the Indian side of Jellalabad, it was not the good fortune of the troops to reach. The road from Khoord Cabul to Tezeen was through a succession of lofty hills, named the Huft Kotul, or Eight Hills.

1842 together. He formed up the 44th and 5th Light Cavalry,
8 Jan. and with these, and the last gun, followed the broken mass to the bivouac at Khoord Cabul, the Affghans being too much occupied with plunder to follow.

The weather chanced to clear up towards evening, or the remainder of the force must here have perished, since it had to undergo, for the third time, the dreadful ordeal of darkness, without fire or shelter. One small tent alone was saved, into which the few wounded that could be brought on with the column were placed. Daylight revealed the miserable plight of the troops, and the morning
9 Jan. of the 9th displayed a sight harrowing in the extreme, the dead bodies of those who had perished during the night being scattered upon the snow in every direction.

Here again communications were opened with Mahomed Akbar Khan, who sent to express his regret at what had occurred in the Khoord-Cabul Pass on the previous day, and promised to afford protection, and forward supplies, if Major-General Elphinstone would halt until arrangements could be made for a safe escort. He also proposed that the married people and families should be made over to him as hostages, promising honourable treatment to the ladies: with this the Major-General complied, being desirous to remove the ladies and children, after the horrors they had already witnessed, from the further dangers of the camp, and hoping that, as from the very commencement of negotiations the Sirdar had shown the greatest anxiety to have the married people as hostages, this mark of trust might elicit a kindly feeling in him. No supplies were, however, received, the sole object of the Affghan chief being to detain the British as long as possible, so as to exhaust their scanty stock of provisions and leave them to be destroyed by cold, famine, and the sword.

The second Shah's cavalry here deserted to the enemy in a body, giving as a reason that as their commanding officer had quitted the corps to seek the protection of the Sirdar, they had no other resource.

Another dreadful night closed in, and when day

dawned on the morning of the 10th, the picture of 1842
wretchedness that had already presented itself on the 10 Jan.
previous mornings was still further heightened in all its gloomy effects. The frost had made sad ravages amongst the Sepoys and camp followers, whose corpses lay stretched around, and the living, almost reduced to despair, began to envy those whose miseries were thus early terminated. The native force was reduced to a shadow, and the few Sepoys who remained faithful to their corps were quite disabled by the cold. This day they formed the rear-guard.

About eight o'clock in the morning the force moved forward, and arrived at the Tungee Pass without other molestation than an occasional shot. As the advance passed through, consisting of the remains of the 44th and European artillery, upon whom now fell the brunt of defending the column, the Affghans were seen collecting within the walls of an old fort a short distance in front, whereupon Brigadier Shelton ordered out a party of the 44th to take possession of the place and drive the people away. This order was countermanded by Major-General Elphinstone, as they appeared peaceably disposed, some of them motioning for the soldiers to continue on, and holding up their hands in token of their friendly intentions. No sooner, however was the advance out of sight, and the baggage and unarmed camp followers before them, than they opened a most destructive fire, killing several, and plundering nearly all the remaining baggage. All the Sepoys and Irregular Cavalry had either quitted the force between this place and Khoord Cabul, or throwing away their arms and accoutrements, had become absorbed among the camp followers: none were observed afterwards. The 5th Light Cavalry, together with the remnant of the 44th, some fifty artillerymen, and one gun, were all that now remained of the force, but the followers still mustered in large numbers.

Brigadier Shelton now took command of the 44th, and, sending one company, under Lieutenant Souter, in advance, with the remainder he formed a rear-guard—the

1842 post of danger—for plunder being the enemy's chief object
10 Jan. all his attacks were directed against the rear, seldom molesting the advance. The force had not proceeded far from the last scene of carnage, when a large body of Affghans was observed advancing; the 44th was formed up on a rising ground, and the gun was placed in position. Their leader proved to be Akbar Khan; and Captain Skinner was sent to him by Major-General Elphinstone to complain of the late aggression. The chief again expressed his regret at not being able to control the Ghilzyes, and said that, although he could not save the natives, yet he would protect the Europeans, on their delivering up their arms, which latter measure he considered necessary to inspire his people with confidence. This was, of course, not acceded to, and he then sent forward a few of his followers as a protecting escort. As the British pursued their weary way, numbers of Affghans, both horse and foot, were seen pushing a-head over the hills to the right and left to the Tezeen Pass. This is a most formidable defile, shut in by commanding heights, from which the enemy assailed the troops on both flanks, at the same time pressing hard upon the rear. It was a severe struggle, and on two occasions they rushed on in such force, and with so much daring, that they were with difficulty repulsed. Had the men but once lost confidence, all must inevitably have then fallen. Sustained by the example of Brigadier-General Shelton, the 44th halted and faced the enemy, and averted, for a time, the impending destruction.

Numbers of officers and men were killed and wounded in this fearful defile: Major Scott, Captain Leighton, and Lieutenant White, of the 44th, the latter of whom had on a former occasion been wounded, were amongst the killed. Upon reaching the Tezeen valley the temperature became considerably milder, and the snow appeared only in patches. Communications were again opened with Akbar Khan, who replied that the arms of the men not having been delivered over, he could not offer any protection, especially as the troops were about to enter a country over which he had no control. What advantage had been derived from his professed aid, even where he had control, may be seen in

the foregoing pages. His professions were only made to 1842
give a pretext for delaying the little army till swarms of 10 Jan.
Ghilzyes could be concentrated on the line of retreat. Had the force reached this place in two days, instead of halting at Bygram and Khoord Cabul, it is possible the greater part might have been able to reach Jellalabad; but the situation was throughout one of great anxiety, for there were in these difficult defiles other enemies to contend with, cold and famine being hardly less destructive than the sword of the merciless foe, whom it had been hoped to elude by negotiation.

A night march being resolved on, it became necessary to leave the last gun, for the horses were completely knocked up, and the valley, for want of a road, impracticable for it in the dark. The march was, accordingly, resumed about seven o'clock in the evening, at dark, when the camp followers, many of whom had large sums of money about their persons, crowded so much upon the remnant of the troops, that it became difficult to prevent their breaking the ranks, for as the enemy's shots came from front or rear, they and the horsemen rushed backwards and forwards like a flood and ebb-tide.

On approaching Burik-ab, about eleven at night, the valley in front was observed to be full of fires, which were naturally supposed to be those of the enemy. The troops advanced with caution, and it appeared strange that not a shot was fired upon them; the mystery was explained on their arriving in the midst of what was thought to be the enemy's encampment, when to their surprise, they came upon the Sepoys, who, having thrown away their arms and deserted, had, by a shorter road, arrived first, and were quietly warming themselves at their fires, not one attempting to move, or offering to proceed with the troops. The Affghans, as has been shown, seldom hurt or molested the Sepoys beyond taking their arms and stripping them of their clothes, so strong is the natural propensity of these Ghilzyes for plunder.

Passing these runaways, the troops were shortly afterwards assailed by rather a sharp fire; this forced Brigadier

1842 10 Jan. Shelton, when about to quit the valley, to form up and return the fire, which had the desired effect of keeping the enemy at a greater distance. The brave little rear-guard executed all orders most intrepidly. Just as day was about to dawn, a party of Affghans from a conical hill close on the left of the road, poured in a volley with a loud shout, but the men of the 44th taking a cool aim, and the figures of the enemy being distinctly visible against the now lightening sky, returned the volley with telling effect.

11 Jan. When daylight at length broke, on the 11th of January, the enemy's fire from all the neighbouring heights was such, that almost every instant some of the poor wearied soldiers were either killed or wounded. Brigadier Shelton never abandoned one wounded man whom it was possible to save—always keeping some of the 5th Cavalry ready to pick up every one that was hit and still able to be moved. Those left dead were cut up with the large knives of their relentless adversaries.

At length a halt took place at Kutter Sung; here there was no water, so after a brief interval the march was resumed for about three miles, and then another halt was made, but still no water was to be had. Meanwhile the enemy, rapidly increasing in numbers, could be seen crowding every height, and his fire became more and more galling.

Major-General Elphinstone took up a position on a height within the walls of an old ruined fort, where he arrived with the camp followers, some of the remaining Artillerymen and 5th Light Cavalry, and the advanced party of the 44th, a considerable time before Brigadier Shelton, who had been detained in repelling the repeated attacks of the Affghans, and in bringing off the wounded.

The road was strewed with dead and wounded, and the enemy's horse having threatened an attack, some of the 5th Light Cavalry and Artillerymen were sent to the rear to keep them in check. So great now was the force of the enemy, that the Brigadier, who had a narrow escape of being captured, was compelled to recall the party. The conduct of the few men of his regiment now remaining, was most noble; they fought with fierce tenacity against the terrific odds opposed to them, and so conspicuous was their gallantry, that those already assembled within the ruined fort, cheered their comrades in this desperate struggle. Lieutenant Frederick J. Campbell Fortye, of the 44th, who had previously been wounded, was here killed. 1842 11 Jan.

The little band of heroes composing the rear-guard, which had by this time become in reality the main body, was now well nigh exhausted. Exposed for six days and five nights on the snow, without cover or fire, with scarcely any food, their feet tender from hard marching or swollen from the effects of the cold, and incessantly harassed by the constant attacks of the enemy, the men were in a most pitiable condition. They were now nearly starved, when two small bullocks, which some of the camp followers had providentially brought thus far, were killed, and instantly devoured raw by the famished soldiers.

Before the men had time to finish this scanty meal, the Ghilzyes had crept up the side of the hill unperceived, and suddenly pouring in a destructive fire upon the unhappy remnant of the force, thus huddled together in a mass, threw the whole into confusion. Forgetful of their fatigue, the men of the 44th flew to arms, rushed at the intruders, and drove them in gallant style off the hill. In addition to their other wants there was now a lack of ammunition, all that remained being in their pouches; this compelled the men to reserve their fire, for they could not afford to return a shot except in cases of extreme necessity, or when their opponents became too daring. As no water could be obtained, the men ate the snow, which only aggravated their thirst. Brigadier Shelton having sent Captain Skinner to the General to request him to take steps to stop the fire of the Ghilzyes, so that his men might move down to the water, this officer, on his return, brought a message to the Brigadier directing him to accompany the General to a conference with Akbar Khan, who was said to be close by, and who had already several officers and all the ladies (including the wife of Sir Robert Sale) and their children with him. It was dark before they reached the Sirdar, who received them

1842 civilly, and as it was understood to be dangerous to return
11 Jan. at night, he was asked to send information of their being detained; but the excuse made was that he was afraid to let any of his people go, one of them having been wounded by the Ghilzyes, now scattered in every direction, and another fired at since darkness had set in.

12 Jan. Thus were the General and Brigadier detained that night and the following day (the 12th), the ostensible pretext being an agreement with the Ghilzy chiefs for a safe conduct through their country; at the same time they were assured that the troops were being supplied with everything they required, while in reality the poor fellows were famishing, and the dastardly Ghilzyes were barricading their line of retreat, and compassing their final destruction. The continued absence of Major-General Elphinstone and Brigadier Shelton, and the more than menacing attitude of the enemy, determined the troops to continue their retreat after dark on the 12th of January; and the night had not long set in when the prisoners, who were only about a mile distant, could hear the firing of the ill-fated remnant, then on the march, and closely pursued by the Ghilzyes. The 44th were now reduced to about eighty men. Before two miles had been passed, after quitting Jugdulluck, the soldiers were stopped by a couple of formidable barriers, which had been thrown up across the road at a narrow pass during the halt. Here all the weak and wounded fell beneath the Affghan knife, and several officers, including Brigadier Anquetil, met their deaths.* After passing these barriers, several of the

* The officers of the 44th who fell at Jugdulluck, and in the Pass, were Lieutenants William Henry Dodgin and Francis Montresor Wade, Paymaster Thomas Bourke, Quartermaster Richard R. Halahan, and Surgeon John Harcourt.

Paymaster Bourke had been nearly forty years in the service, which he entered as Paymaster in 1804. He had joined the 44th in 1823, and served with the regiment in Arracan. Some of the officers of the avenging army recognised the remains of the poor old man, from there being a small portion of his silvery grey hair still adhering to the skull. Many valuable papers were lost with his effects; the funds of the regiment, which were unusually flourishing, were in his hands, and some of them were altogether lost. What appeared to be a piece of dirty paper was picked up in the Tezeen

officers of the force, who were mounted, left the troops and pushed on by themselves; of these Assistant-Surgeon Brydon, of the Indian Army, alone succeeded in reaching Jellalabad. Others remained and continued on during the night with the remnant of the 44th. This handful of men, now only half the number that had started from Jugdulluck, were again fiercely assailed while crossing the Soorkaub river, where Lieutenant Henry Cadett, of the 44th, was shot. 1842 12 Jan.

Shortly after daylight, on the 13th of January, the exhausted survivors found their progress arrested by a numerous body of horse and foot, in a strong position across the road, whereupon they ascended a height on their left hand, and, reaching the top, waved a handkerchief; some of the Affghans then came to them and agreed that 13 Jan.

Valley, and proved to be an order for £300, belonging to the officers' mess fund. The amount was recovered by the regiment.

Quartermaster Halahan had been a lieutenant in the 80th Regiment, but was placed on half-pay on the reduction of the army in 1817. He was appointed Quartermaster of the 44th in 1822, and served with the regiment in Arracan. He was of great strength and known to be the most powerful man in the regiment. He carried a musket from Cabul, and fought in the ranks, killing many of the enemy, his face being black with powder. He had been wounded in Cabul, at the Commissariat Fort, and fell while crossing the barrier in the Jugdulluck Pass.

Lieutenant Dodgin had lost a leg near Peshawur, when on the march to Cabul, in the following unlucky manner:—He was at tiffin in his tent with Quartermaster Halahan, when a cry was raised in the camp of "a man running a muck." Dodgin stepped out to see; and it turned out to be a Syce he had discharged that morning, who was making straight for the tent, brandishing a sword as sharp as a razor. Dodgin called to Halahan, who came out with a thick stick, and felled the man lifeless with a single blow; but not in time, however, to aid poor Dodgin, who, in attempting to step out of the fellow's way, stumbled over a tent-rope, and received from him so severe a wound as to occasion amputation of the leg. Lieutenant Dodgin also was killed at the barrier in the Jugdulluck Pass. His name was subsequently gazetted to a company in the 44th, "vice Robinson," his promotion being ante-dated 15th November, 1841.

1842 13 Jan. Major Griffiths (37th Native Infantry) should proceed to the chief of Gundamuck to make terms; whilst he was gone, a few of them gave the men some bread, and possibly gaining confidence from this, they yielded to their usual propensity for plunder and endeavoured to snatch the arms out of the soldiers' hands, when an officer exclaiming, "Here is treachery!" words came to blows. The Affghans were instantly driven down the hill; firing was then recommenced and continued for nearly two hours, during which the heroic remnant kept the enemy at bay, till their numbers being reduced to about twenty, and their ammunition expended, the Affghans rushed in suddenly with their knives. An awful scene ensued, and ended in the massacre of all except Lieutenant Thomas Alexander Souter, Lance-Sergeant Alexander Fair, and six soldiers of the 44th, three artillerymen, and Major Griffiths, 37th Native Infantry, whose lives the Affghans, with unwonted humanity, spared. In this last struggle Lieutenants Thomas Collins, Arthur Hogg, Edward Sandford Cumberland, Samuel Swinton, and Dr William Primrose, Assistant-Surgeon, all of the ill-fated 44th, were killed.

Thus terminated the last stand made by the 44th, to which the remark of Francis the First is specially appropriate, that "all was lost save honour!" Never, in the military history of Great Britain have her soldiers been called upon to take part in a more trying struggle than this. Never has a nobler resistance been offered to a fate against which human efforts were of no avail.

Of the one hundred and two officers who were killed at Cabul, and during the retreat, twenty-two belonged to the 44th.*

---

* The following interesting letter was received from Colonel Ross Thompson, of the Royal Engineers, in June, 1882, while the regiment was serving at Madras:—"On Saturday, the 10th of January, 1880, I visited the memorial on '44th Hill,' near Gundamuck, and carried away with me the enclosed sprig of a little shrub which grows there—it is somewhat like a myrtle. I think you are aware that a monument was put up to the memory of the 44th by the Madras Sappers (Queen's Own) in 1879—it was knocked about to a certain extent in the interim between the wars of

Of six hundred and eighty-four men of all ranks of 1842
the 44th, at Cabul, on the 1st of October, six hundred and thirty-two had perished; the remainder, with the exception of the seven who survived the last stand at

---

1878–1879, and 1879–1880, so before quitting the country in August 1880, I sent (with General Bright's permission) a party of the Queen's Own Sappers to repair it. The '44th Hill' was a place of deep interest to us all. An old Afghan Khan, living near Gundermuck, told us that never did men make a more gallant and desperate stand against overwhelming odds than was made by your old corps. *He was present at the fight and their annihilation.* Of course, as an Afghan, he rejoiced in seeing the invaders of his country fall, but he was a gentleman, and paid all possible respect to their gallantry and memory."

List of officers and their fates :—

Lieut.-Colonel J. Shelton, prisoner, 11th January, 1842.

Lieut.-Colonel T. Mackrell, died of wounds received capturing Rika-bashee Fort, 10th November, 1841.

Major W. B. Scott, wounded, 8th January, 1842, Khoord Cabul Pass; killed, 10th January, 1842, Tezeen Pass.

Captain Thos. Swayne, killed, 4th November, 1841, Commissariat Godown.

Captain R. B. M'Crea, killed, 10th November, 1841, Rika-bashee Fort.

Captain T. R. Leighton, wounded, 8th January, 1842, Khoord Cabul Pass; killed 10th January, 1842, Tungee.

Captain T. Robinson, killed, 4th November, 1841, Commissariat Godown.

Lieutenant W. H. Dodgin, killed, 12th January, 1842, at Barrier.

Lieutenant T. Collins, wounded, 11th January, 1842, Kutter Sung; killed at Gundamuck.

Lieutenant W. Evans, left in charge of sick at Cabul.

Lieutenant W. G. White, wounded, 7th January, 1842, Bootkhak; killed, 10th January, 1842, Khiela Thanak.

Lieutenant T. A. Souter, wounded and taken prisoner, 13th January, 1842, at Gundamuck Hill.

Lieutenant F. M. Wade, killed, 12th January, 1842, Jugdulluck Hill.

Lieutenant A. Hogg, wounded, 4th November, 1841, Commissariat Godown; killed 13th January, 1842, Gundamuck Hill.

Lieutenant E. S. Cumberland, wounded, 5th November, 1841; killed 13th of January, 1842, Gundamuck.

Lieutenant W. G. Raban, killed, 6th November, 1841, capturing Mahomed Shereef.

Lieutenant H. Cadett, killed, 12th January, 1842, at Soorkaub.

Lieutenant S. Swinton, wounded, 23rd November, 1841, Be-maroo Hill; killed, 13th January, 1842, Gundamuck.

1842 Gundamuck,* and nine, who had been taken prisoners during the retreat, had been left at Cabul sick or wounded, and of these fourteen died while in captivity.

It is a telling fact for the 44th, that they all died fighting to the last, with arms in their hands, as soldiers, and under a pressure of famine, cold, and fatigue, which has no parallel for privation and suffering. The numerous skeletons found six months afterwards by Major-General Pollock's force, in circles and heaps at Gundamuck, Jugdulluck, and other halting-places and in twos and threes at short intervals between, plainly indicated the route of the doomed force, and told the sad history of its dying struggle and unavailing heroism.

While Major-General Elphinstone and Brigadier Shelton were endeavouring to arrange with Akbar Khan at Jugdulluck, and on the troops being suddenly attacked, Lieutenant Souter, seeing the perilous situation of the colours, tore one (the Regimental) from the staff, and wrapped it about his body. Lieutenant Cumberland took the other, the Queen's, but not being able to button his coat over it, handed it to Colour-Serjeant Patrick Carey, of the grenadier company, who secured it about him, under

---

Lieutenant F. J. C. Fortye, wounded, 3rd November, 1841, at Mahomed Shereef; killed, 11th January, 1842, entering Jugdulluck.

Ensign A. W. Gray, wounded, 6th December, 1841, at Mahomed Shereef; killed between Gundamuck and Neemlah.

Paymaster T. Bourke, wounded and died at Jugdulluck, 12th January, 1842.

Quartermaster R. R. Halahan, wounded, 4th November, 1814, garden adjoining Commissariat Godown; killed, 12th January, 1842, at Barrier.

Surgeon J. Harcourt, killed, 12th January, 1842, at Barrier.

Assistant-Surgeon W. Balfour, killed, between Gundamuck and Neemlah.

Assistant-Surgeon W. Primrose, M.D., killed, 13th January, 1842, Gundamuck Hill.

* One of these survivors, Private Moore, 44th, who acted as his servant, saved, for a time, Major-General Elphinstone's life. On the earthquake occurring, which nearly destroyed Sir Robert Sale's fortifications at Jellalabad, he carried the General, who was bed-

his posteen;* this non-commissioned officer was killed during the night of the last march, and amid the snow, confusion, and darkness, and the enemy pressing closely upon them, the men failed in their endeavours to recover it from his body, and the Queen's colour was thus unavoidably lost. 1842

Lieutenant Souter, in a letter to his wife, from his captivity near Lughman, in the hills, not many miles from Jellalabad, thus wrote :—" In the conflict my posteen flew " open and exposed the colour. They thought I was some " great man, looking so flash. I was seized by two fellows " (after my sword had dropped from my hand by a severe " cut in the shoulder, and my pistols had missed fire) ; " they hurried me from the spot to a distance, took my " clothes from off me except my trowsers and cap, led me " away to a village by command of some horsemen that " were on the road, and I was made over to the head man " of the village, who treated me well, and had my wound " attended to. Here I remained a month, seeing occasion- " ally a couple of men of my regiment who were detained " in an adjoining village. At the end of a month I was " handed over to Akbar Khan, and joined the ladies and " the other officers at Lughman. I lost everything I " possessed. * * * * * * My wound, which is " from my right shoulder a long way down my blade bone, " is an ugly one, but it is quite healed. The cut was " made through a sheepskin posteen under which the " colour was concealed, lying over my right shoulder, that " thick Petersham coat I used to wear at Kurnaul, a flannel " and shirt. My sword fell from my hand, and the pistol " I had in my left hand missed fire. I threw it then upon " the ground, and gave myself up to be butchered. The " man I tried to shoot seized me, assisted by his son-in-law, " and dragged me down the hill; then took my clothes, the " colour, and my money. I was eventually walked off to a " village two miles away. This same man and his son-in-

---

ridden and unable to move, in his arms from the house, which, in a minute afterwards, was a crumbling heap of ruins from the shock.

* Sheepskin coat, worn by the troops.

1842 " law, whose names are Meer Jann, came afterwards to the " village where I was, with my telescope, to get me to show " them how to use it. Afterwards the son-in-law and I " became thick : he brought me back the colour (though " divested of the tassels and most of the tinsel) one day, to " my agreeable surprise." Both the colours had been for some years but a mere bundle of ribands.*

---

* The following is a copy of a letter now in the possession of Akram Khan, of Tutu, son of Shamshadin Khan, written in ink :—

"MY DEAR FENWICK,—To my sorrow, the kind hopes entertained in your letter that I should soon be released have not been realized, my detainer having increased his demand for my ransom. I have written a long letter to Captain MacGregor by my detainer, who takes it to Brigamer Fort (illegible),—pray go to MacGregor and read my letter to him, and try what can be done to obtain my release. My family must have heard that my regiment is destroyed, and are naturally in great distress about me. I have recovered the regimental colour of my regiment, which I had about me when I was cut down and taken prisoner. The other colour was in the possession of a sergeant, who was killed upon the road during the night. There are two men belonging to my regiment in this neighbourhood recovered from their wounds. An Artilleryman died three days ago.

" Yours sincerely,
" C. A. SOUTER.

" Addressed to Captain FENWICK,
13th Regiment, Jellalabad."

(On the back, in pencil.) "I write this on, I believe to be Sunday, the 30th January, 1842. The Malick, though he promises to start to-morrow, I apprehend it will be the following day before he does do so. I have neither seen, nor have I been able to hold any intercourse with Major Griffiths since the first day I arrived, now nearly three weeks, a long time to wear a bloody shirt."

N.B.—The above letter was shown to the undersigned, by whom the copy was made, by Akram Khan, who states that he found it on the Cabul high road, where no doubt the bearer had thrown it down instead of taking it to Jellalabad. Akram Khan was then a boy, and says that "Souter Sahib" lived in his father's house.

True copy.
(Sd.) C. STRAHAN, Capt., R.E.,
Dep. Sup. Survey of India.

Safed Sang,
May 19th, 1879.

Brigadier Shelton, in writing, stated, with regard to 1842
the treatment of the prisoners: "On the 15th (January) "we proceeded to Tiggeree, where Lamech, the father of "Noah, was buried, and on the 17th to this place "(Lughman), where we met with every attention to our "wants: in truth, the Sirdar has treated us with much "kindness ever since we have been under his protection, "nor have the ladies cause to complain, he having sent "them quantities of cotton, chintzes, and long cloth for "themselves and their children."

No sooner had intelligence reached India of these untoward events than it was resolved to relieve Jellalabad, then being defended by the gallant garrison under Sir Robert Sale, who, at the time of the outbreak at Cabul, was employed in clearing the Khoord Cabul Pass, and had subsequently marched on Jellalabad; this place was reached by Major-General (afterwards Sir George) Pollock in April, the Affghans having then abandoned the siege, after their repulse on the 7th of that month by the defenders of the fortress. Here, at this time, were consigned to their last resting-place, with military honours, the remains of the unfortunate Major-General Elphinstone, who died whilst in the power of Akbar Khan, and whose body was sent by that chief to Jellalabad for interment.*

Some hard fighting took place in the neighbourhood after the arrival of the force from India, particularly with the warlike and turbulent Shinwarree tribes, who had attacked several convoys and been guilty of many acts of murder and plunder. This service occupied some time and was not completed until towards the end of July.

All were now naturally anxious to move on Cabul, for which city the army, in two divisions, proceeded on the 20th of August. On reaching the Pass of Jugdulluck the bleached remains of their comrades were seen by the advancing troops, and after a succession of day and night skirmishes, the Affghans were so completely defeated on

* Private Moore, 44th, was allowed to accompany the remains of his late master; the escort was attacked by Ghilzyes, and he was wounded.

1842 the 13th of September, that Akbar Khan, with a solitary horseman, had great difficulty in effecting an escape from the field. Two days afterwards the British colours were hoisted on the highest points of the battlements of the Balla Hissar, where they continued to be displayed during the stay of the force at Cabul.

Major-General (afterwards Sir William) Nott having advanced from Candahar, and recaptured Ghuznee, effected a junction with the troops at Cabul on the 17th of September. Four days afterwards the release of the prisoners was effected: Lady MacNaghten, and the wife of Sir Robert Sale being amongst the number, in all one hundred and five. The following belonged to the 44th Regiment:—

Brigadier-General Shelton; Lieutenants Evans and Souter;* Colour-Sergeant James Wedlock, Sergeant James Weir, and Lance-Sergeant Alexander Fair;† Corporals, Thomas Beavan and William Sumpter; Drummers, Thomas Branigan (died 8th November following), John Higgins, and Thomas Lavall; Privates, William Arch, Patrick Burns, Patrick Brady (died 21st November), Michael Cresham. Timothy Cronan, Robert Cox, Daniel Driscoll, Martin Devany, Patrick Duffy (died 2nd December), Elisha Durrant, James Mathews, John McDede, Daniel McCarthy, James McCabe, Thomas M'Glynn, Patrick Marron, William Moore, James Miller, Patrick Murphy, John Marshall, James Nowlan, Errington Robson, James Robinson, James Seybourne, Jeremiah Sheehan, John Stott (died 20th October), William Tonge, and George Wilson; Boys, William Grieas and Joseph Millwood.

George Reynolds, Orderly Room Clerk; Corporals, Robert Rodgers and Henry Smith; Drummer, John

---

* Lieutenants Evans and Souter were promoted to companies in the 44th, on the 14th of October following. They afterwards exchanged into the 18th and 22nd Regiments respectfully.

† Lance-Sergeant Alexander Fair was transferred to the 50th Regiment, with which he served in the Sutlej Campaign of 1846. He was promoted Quartermaster in that regiment in 1848, and subsequently Paymaster.

Hamlinton; Privates, Daniel Collins, William Connors, 1842
Barnard Handley, Charles Kennedy, Joseph Lewis, Dennis Murray, Michael O'Brien, Timothy Rock, Cornelius Tierney, and James Wackerley, died during their captivity.

All the objects of the campaign being accomplished, orders were given for the evacuation of the country, and the troops commenced their march to India on the 12th of October.

Much undeserved obloquy has been cast on Brigadier Shelton, Lieut-Colonel of the 44th, and it is therefore due to his memory to record that, Major-General Elphinstone, in a letter to the Secretary of Government, in the Secret Political Department, Calcutta, from Buddeeabad, dated 23rd of February, 1842, expressed his "Sense of the gallant "manner in which the various detachments sent out were "led by Brigadier Shelton, and of the invariably noble "conduct of the officers on those occasions, particularly "those who fell leading their men, namely,—Colonel "Mackrell; Captains Swayne, Robinson, M'Crea; and "Lieutenant Raban, of Her Majesty's 44th Foot."

In confirmation of this praise it may be added that 1843
the finding of the court-martial, assembled for his trial on the 20th of January, 1843, at Loodianah, was that this officer "gave the orders for preparations for a retreat from "Affghanistan, but not prematurely, nor without authority, "nor in the absence of instructions from the chief author-"ities;" and in their remarks they "expressed their "conviction that Brigadier-General Shelton was placed "under circumstances at Cabul of a most unexpected, "unusual, difficult and distressing nature; and that the "evidence and documents before the Court exhibit proof, "on his part, of very considerable exertion in his arduous "position, of personal gallantry of the highest kind, and "of noble devotion as a soldier.*

* Colonel Shelton was a meritorious officer, a brave soldier, and a well-informed man, and was the only fighting General at Cabul. He had seen hard and distinguished service in the 9th foot, but was not popular. The Duke of Wellington, nevertheless, was so well

1843 And the remark of General Sir Jasper Nicolls, K.C.B., Commander-in-Chief, in the East Indies, upon the fourth charge, namely—" For having at Jugdulluck, on the 12th " of January, 1842, during the retreat of the British Force " from Cabul towards Hindoostan, suffered himself to be " taken prisoner by the enemy, by want of due precaution," of which he was fully and honourably acquitted, was " that his acquittal is most satisfactory and complete."

---

satisfied with his conduct, that he entrusted to him the re-formation of the regiment. Shelton had been in three eventful retreats—that to Corunna, that from Burgos, when he was in the 9th Foot; and that from Cabul. Having only one arm, he must have been the personification of suffering and endurance whilst battling in vain against overpowering numbers.

The following letter by "An old 44th" appeared in the "Daily News" newspaper:—

"In a recent article in your columns upon the Cabul Massacre, the following passage appears, viz.:—'Sale's example establishes the truth that the calamity which has befallen our soldiers in Cabul might have been avoided if wiser counsels and more resolute policy had been adopted.' And again, 'who can reasonably doubt that what Sale had done at Jellalabad, Elphinstone, or some Commander less enfeebled by ill-health, could have accomplished at Cabul?' In justice to the memory of the late Colonel John Shelton, who was then commanding the 44th Regiment, and 2nd in command in Cabul, I deem it right to mention what did occur at a Council of War held some short time previously to the evacuation. Colonel Shelton had a habit of cutting any piece of wood that was near him with a penknife, and, whilst the conference was proceeding, sat silent, making no remark, but shaving off little slips from the table. This annoyed the envoy, Sir William Macnaghten, who sharply remarked, 'I think at a time like this, Colonel Shelton, you might be doing something better than whittling sticks. Pray, Sir, will you give us your advice?' 'My advice,' replied Colonel Shelton, 'is that General Elphinstone should at once place himself upon the sick list, for he is at this moment so ill that he is utterly unfit for duty, and ought to be in his bed, and having done so, place me in command.' 'Well, Sir,' said the envoy, 'and if in command, what would you then do?' 'I should at once,' replied Colonel Shelton, 'place the Bala Hissar in such a state of defence as to enable us to stand a siege; I should make sorties, and bring in cattle, and grain, and wood, and send down to India for an army to come up and relieve us.' 'And you would withdraw all protection from Shah Soojah?' retorted the envoy. 'No,' said Shelton, 'let him stay in our midst and we will protect him from all the Afghans; but I tell you that

A detachment of the 44th (three hundred and twenty-seven strong), under Brevet-Major Johnston, had proceeded from Cawnpore in November, 1841, for Affghanistan, and on marching through Delhi in December, heard, for the first time, of the Cabul outbreak. The detachment was afterwards attached to Major-General McCaskill's brigade, and did duty and paraded with the 9th Foot, of which he was Lieut.-Colonel, on the advance across the Sutlej of the relieving force under Major-General Pollock. After the detachment had proceeded some two or three marches into the Punjaub, and when at Kussoor in camp, at eleven o'clock at night, an orderly from head-quarters brought the order for its return at once to India, the words of Sir Jasper Nicolls being—"the detachment of the 44th now "proceeding with the force to Affghanistan will cease to "belong to it from the receipt of this order, and will fall "back by forced march to Ferozepore, as it goes to join a "regiment *which does not exist.*" This was in January, 1842; and the detachment had the mortification, on leaving the force the following morning, of seeing the several regiments marching to avenge their slaughtered brethren. During its subsequent stay at Ferozepore, where the detachment was encamped for five months, one hundred 1843

---

once you leave here you will not be able to protect yourselves.' 'I could sanction no such act,' replied the envoy, 'or consent to such a step.' 'Well, Sir,' said Colonel Shelton, 'that is my opinion and advice, and if you put me in command that is what I should do: as it is, give me whatever orders you and General Elphinstone think fit and I shall carry them out to the best of my ability.' Alas, his advice was rejected, and all who left Cabul on the final retreat were massacred, except Lieutenant Souter and seven soldiers of the 44th and three European Artillerymen, with one company's officer, whose name I forget. Colonel John Shelton was not a pleasant man upon parade, nor the most popular commanding officer in the Army; but he was a gallant soldier, a brave and determined man. He had seen service and knew what war was, for he fought at Roleia, Vimiera, Corunna, Flushing, Badajoz, Salamanca, Madrid, Burgos, Vittoria, St. Sebastian (where he lost an arm), Ava and Arracan, and a host of minor battles. This is a slight record, due to the memory of Jack Shelton, from one that feels he deserves it, and who had the honour of serving as an officer of the 44th in India, when he was at its head."

1843 and fifty of the three hundred and twenty-seven men died from asphyxia. The detachment was afterwards moved to Kurnaul.*

This portion of the 44th, increased by the arrival of two drafts from home, after giving volunteers to other corps continuing in India, embarked for England early in February, 1843.

On the 7th April, 1843, Lieut.-General the Honourable Patrick Stuart, colonel commandant of the second battalion of the 60th Regiment, was appointed to the colonelcy of the 44th, in consequence of the decease of General Gore Browne, on the 12th January.

Great efforts were made to complete the regiment, and during the year it received six hundred and fifty-seven recruits. The depôt, now become the regiment, thus rapidly increasing in strength, was moved from Chatham to Canterbury, and thence in a few weeks to Deal, at which place, on the 27th of May, it was formed into ten companies. The men from India, two hundred and twenty-one in number, joined here in June. In July the regiment was removed to Chichester, and in August it proceeded to Gosport.

Whilst the regiment was stationed at Gosport, Colonel Burney—whose reminiscences have afforded much additional information for this Record—went to see his old friend Colonel Shelton, at that time quite a wreck of his former self. As they were conversing on by-gone times, the colour of the 44th, which stood in a corner near which

---

* Whilst quartered here Lieutenant P. W. MacMahon, Adjutant and Interpreter of this detachment, collected the testimony of many of the Sepoys and camp-followers who were daily arriving from the disastrous retreat, as also copies of Major-General Elphinstone's despatch, and of letters from Colonel Shelton and Lieutenant Souter. These, as carefully arranged by him, elicited at the time the thanks of Sir Jasper Nicolls, then Commander-in-Chief in India, and are the groundwork of the foregoing account.

Colonel Burney was sitting, fell down and struck him on the head; this was considered an extraordinary circumstance, especially by his brother, a sailor, upon whom it made a great impression, and who could not forget the reception given by the colour to its former officer. 1843

In October new percussion muskets were issued to the regiment. On the 1st of December the strength of the 44th amounted to thirty-four sergeants, fourteen drummers, and eight hundred and eighty-two rank and file.

At Haslar, on the 11th December, new colours were presented by Lady Pakenham, wife of Major-General the Honourable Sir Hercules Pakenham, K.C.B., who spoke as follows:—" In presenting the colours of the 44th Regi- " ment, I am deeply moved, for I cannot help feeling a " recurrence of those painful emotions their late history in " Affghanistan excited when perusing it; yet when I look " around me on the few brave men who survived those " almost unheard-of perils, and those who have since joined " their ranks, I feel assured that the future career of the " 44th Regiment will only remind the world of its former " existence by displaying to more than ordinary advantage " the well-known bravery of British soldiers, and so—if I " may be allowed an allusion to the fabled bird of old— " rise like a phœnix from the ashes of its predecessor. " Receive these colours with the warmest prayers for their " future glory and success from one who, as the sister, " wife, and mother of soldiers, feels an enthusiastic interest " in that brave order of men. May they always be unfurled " in support of the honour of our country and our gracious " Queen, and to the glory of the King of Kings, whose " blessing has already been so eloquently invoked."

The new colours were consecrated, with a most impressive prayer and address, by Archdeacon Wilberforce (afterwards Bishop of Oxford), then rector of that parish.

Colonel Shelton, after returning thanks, added— " I must now pay a tribute to the memory of the departed " brave, by vindicating the aspersions which have been

1843 " so ungenerously cast upon the conduct of a part of the " regiment. When the Sepoys deserted us, and the " Europeans were left to their own resources, the men of " the 44th showed their sterling worth. From that " moment I assumed the command of the rear-guard; " and in going through the Tezeen Pass—annoyed by a " galling and destructive fire from the heights on both " flanks, and when crowds of savage Ghilzyes rushed like " a torrent upon the rear—this brave little band, obedient " to my voice, halted, faced about, and repelled the " appalling numbers of the enemy under a tremendous " fire, with a boldness and determined courage that " might have extracted admiration from the very stones " under their feet; and though they had been then four " days and four nights on the snow, these noble fellows " performed a forced march of two days and one night, " without halting—repelling the incessant attacks of the " enemy, under a destructive fire that strewed the whole " line of road with the dead bodies of their comrades, with " a constancy and unyielding courage that will command " my admiration and respect so long as I live—and what " I now state was under my own personal observation. No " man ever thought of surrender, but all fell gloriously " with their arms in their hands, fighting to the last, and " only sixteen remained of the number that marched from " Cabul.

" I need now merely add that the conduct of the old " regiment is well worthy the emulation of the present " young corps, which has only to show similar devotion " and I shall be proud to lead it against the bravest foe " Europe can boast."

The old colour, saved by Lieutenant Souter, having been trooped, was, with a monument to the memory of the officers and men who fell in Affghanistan, placed in the church of Alverstoke, Hants.

1844 In January, 1844, the regiment was moved from

Gosport to Winchester,* and in May following to Devonport, where it remained until the spring of the next year. 1844

On the 14th of April, 1845, the 44th embarked at Devonport for Ireland, and arrived at Dublin on the 22nd of that month. 1845

Here Colonel John Shelton met with a fatal accident on the 10th of May. His horse fell with him in Richmond Barrack Square, Dublin, and he fractured his skull. He lingered for three days, insensible, and died on the 13th. This was, indeed, a melancholy termination to the life of a soldier who had braved death in many a hard fought battle during a service of forty years.

Colonel Shelton joined the army in 1805, and first served with his regiment—the 9th Foot—in the first campaign in the Peninsula in 1808, under Sir Arthur Wellesley, including the battles of Roleia and Vimiera, and subsequently under Sir John Moore, in the retreat to, and victory at, Corunna. He next served in the Walcheren expedition in 1809, and was present at the siege and capture of Flushing. He served in the Peninsula again under Wellington, and was present at Badajoz, Salamanca, Madrid, retreat from Burgos, and assault and capture of St. Sebastian—at which latter he lost his right arm. It is related of him that he *stood* at the door of his tent, with an unmoved countenance, while the surgeon performed the amputation, which was from the socket, and that he was again under fire with his regiment within three months. He next served in the campaign in Canada in 1814, and in 1817 exchanged into the 44th Regiment, with which he went to India, and served first in the campaign in pestilent Arracan, and after-

* Captain Evans joined here, with Sergeant Weir, Privates Moore, Miller, and McDede. These were the only men who had served in Cabul and the retreat, who joined the regiment after its return home. The two former were discharged, and the two latter died in the regiment shortly afterwards. Pensioner-Sergeant Weir died at Madras in 1859, while the regiment was quartered there. He had proceeded thither from New South Wales, in charge of a cargo of horses.

1845 wards, as brigadier-general, in the ill-fated force in the outbreak in Affghanistan, as recorded in these pages. After his return to cantonments from the Balla Hissar there was not an expedition, or even a sortie, that he did not lead in person, animating, by his fearless bearing, the troops engaged; and in the disastrous retreat he was ever to be found where the attacks of the enemy were fiercest, encouraging the hard-pressed soldiers—scorning danger, and bearing, apparently, a charmed life. The accident, however, which caused his death, was mainly attributable to his loss of an arm, for he was unable, in consequence, although a good rider, to restrain or guide his run-away horse, which slipped and fell with him, on his unprotected side, with such force as to fracture his skull.

It will hardly be credited, in these days of medals, that Colonel Shelton, with all his service, went to his grave without a decoration—the war-medals for the Peninsula and India not having at that time been granted. He was succeeded in the Lieutenant-Colonelcy by Major the Honourable Augustus Almeric Spencer.

1846 From Dublin the regiment proceeded, in March, 1846, *en route* to Newry, where it arrived on the 2nd of April. On the 19th of August following it marched to Belfast, at which station the regiment arrived two days afterwards. In the autumn, orders were received to recruit to twelve hundred rank and file, with a view to the formation of a reserve battalion.

1847 The regiment returned to Newry in the beginning of April, 1847, and proceeded to Fermoy on the 16th of December, at which station it arrived on the 29th of that month. In February of this year the strength of the regiment reached one thousand two hundred and twenty-seven rank and file, and in the following month it was divided into twelve companies.

1848 On the 18th of February, 1848, the 44th marched from Fermoy to Cork; and, having been ordered for foreign service, was, on the 1st of April, organised into two battalions—the first of which embarked for Malta on the 11th

of April, and arrived on the 8th of May. The other—the 1848
reserve battalion—embarked for the same destination, under Major H. O. Moore, on the 20th of June, and arrived on the 14th of July. The depôt was moved to Parkhurst, Isle of Wight. Lieut.-Colonel J. O. Clunie, C. B.—who had been appointed to the command of the reserve battalion on its formation, from half-pay of the Third Buffs—was unable to embark with it in consequence of ill-health. He retired in November, 1848, and was succeeded by Major A. H. Ferryman, who exchanged, on the 6th of September, 1849, with Lieut.-Colonel E. Thorpe, of the 89th Regiment.

During the years 1849 and 1850, the 44th remained at 1849
Malta—the depôt continuing at Parkhurst. In April, 1850
1850, the regiment was reduced to ten companies—a thousand rank and file in all, and the reserve battalion was accordingly broken up and consolidated with the first battalion, under Lieut.-Colonel Hon. A. Spencer. The depôt was removed to Chatham. Cholera raged at Malta during July 1850, when in six weeks, two sergeants and seventy-two rank and file died—Assistant-Surgeon Gray falling a victim to this fatal disease, Of the above number, Captain Carey's company (K) lost forty-two out of its total strength of ninety-five men. The regiment also lost fourteen women and as many children, chiefly from the wing in the Cottonera District, where the families were quartered in the same barracks as the above company. The greater number of all these deaths occurred in the last ten days of July.

On the 25th of March, and 1st and 6th of May, 1851, 1851
the regiment, in three divisions, embarked for Gibraltar, and landed at that station on the 19th of April and 20th of May.

The regiment remained at Gibraltar during 1852* and 1852
1853, and until the spring of the following year. 1853

* On the 10th of July, 1852, (two days after his promotion), the regiment sustained a great loss in the death of Lieutenant and Adjutant John Colpoys, who, in addition to his efficiency as an officer

1854 Events of momentous importance had occurred in Europe, and after the celebrated words of Lord John Russell, in the House of Commons, on the 17th of February, 1854—"May God defend the right," England nerved herself for an approaching contest with Russia, whose attempts to crush Turkey had aroused the nation, slow to war, and conscious of the benefit of a prolonged peace, to support the independence of the Ottoman empire.

Lord Raglan, so long associated with the Duke of Wellington, was selected to command the British forces ordered to the East—Malta being the first place of rendezvous. On this service the 44th embarked at Gibraltar, on the 10th of March, and in *The London Gazette* of the 28th of that month appeared the official declaration of war against Russia, and that of the Emperor of the French; that nation being, not as formerly the foe, but now the ally of England. More than half a century had elapsed since the 44th and other British regiments had fought against France in aid of the Sultan, and now, for the first time since the Crusades, French and English troops were leagued in a common cause. The 44th arrived at Malta on the 14th of March.

On its embarkation for service, the total strength of the regiment consisted of eight companies, comprising thirty officers and nine hundred and twenty-three non-commissioned officers and men, leaving a small nucleus for two companies (of sick men, &c.) to be sent to the depôt, which was removed from Chatham to Walmer. The regiment was a remarkably fine one; no man being of less than three

---

and adjutant, was a most energetic promoter of sports and athletic exercises, in all of which he himself was highly proficient, and in consequence of his exertions, the men of the 44th were always found to excel at meetings for garrison games, &c. This officer may also be said to have originated (with the approval of Mr. Angelo), the bayonet exercise of the army. The regiment was drilled in it by him as early as 1848, and was put through it at inspections before Lieut.-General Ellice, at Malta; and Lieut.-General Sir Robert Gardiner, K.C.B., at Gibraltar; both of whom expressed their high approval. With but slight alterations, this exercise is now a part of the established drill of the army.

years, and few of more than eleven years' service. Since 1854 its formation on return from India, the 44th had been almost the whole time under the command of Lieut.-Colonel Hon. A. A. Spencer, to whom, and the excellent system carried out by him, its high character for conduct, appearance, and discipline, was mainly attributable. Whilst at Malta, two hundred Minié Rifles were issued to the regiment.

The stay of the 44th at Malta was but short, as the force proceeded thence to Gallipoli, for which place it (the second regiment of the expeditionary army) embarked on the 4th of April.

Here the first six regiments of the expedition were landed, and were employed by brigades alternately, and in conjunction with a French division, in the construction of a line of intrenchments across the Isthmus, at Boulair. These regiments (with the exception of the second battalion Rifle Brigade, and 93rd Highlanders, which were moved to Scutari, the rendezvous of the rest of the army, to join the light and first divisions there being formed, and which were relieved by the 1st and 38th from England,) were formed into the third division, under the command of Lieut.-General Sir Richard England, K.C.B.

First brigade under Colonel Sir John Campbell, Bart.—
1st, 38th, and 50th Regiments.

Second brigade under Colonel William Eyre, C.B.—
4th, 28th, and 44th Regiments.

The 44th was thus amongst its companion corps of the old war; having served in the Peninsula in the same brigade with the 4th, and in the same division with the 28th at Waterloo: also in the same brigade with the 1st at Waterloo, and in the same division with the 1st and 38th in the Peninsula.

From Gallipoli, the division was moved to Varna, whither the bulk of the army had preceded it; the 44th embarking on the 23rd and 24th of June, and arriving there three days afterwards. Here the army was assembling, and the third division was encamped in the neighbourhood

1854 of the town until the embarkation of the force for the Crimea. The 44th were fully employed in making fascines and gabions.* There were a few fatal cases of cholera in the regiment, but the light and other divisions suffered more than the third. On the 6th of August, orders were received for the infantry to wear moustachios. Whilst at Varna,† the regiment was completed with rifles of the Minié pattern.

In consequence of the successful defence of Silistria by the Turks, and the repulse of the Russians before that fortress, the fear of Turkey being invaded passed away and as the allied armies were no longer required to support by their presence the effort of the Ottoman troops, it was resolved to make a descent upon some of the possessions of Russia, and eventually the expedition to the Crimea was determined upon.

Accordingly, the whole expedition embarked at Varna. The 44th, under the command of Colonel Hon. A. A. Spencer, on the 29th of August, on board the steam transport, "Tynemouth": its strength was thirty officers, and eight hundred and ninety nine men of all ranks. The names of the officers were—Colonel Hon. A. A. Spencer; Majors Staveley and Fielden; Captains MacMahon, Browne, Fenwick, Faussett, Robinson, Howard, Hon. C. Agar, and

---

* Two hundred men of the regiment, under Major Staveley, were detailed for duty with the cavalry division, to look after Captain Nolan's horses, then arriving in great numbers from Africa, there not being dragoons sufficient to spare for the purpose. The men made themselves so useful, and such expert horsemen, that the Earl of Lucan afterwards applied for the transfer of the whole two hundred permanently to the cavalry, but this was, of course, refused.

† Whilst the army was at Varna, there was a remarkable flight of locusts; they passed in a dense flock over the camp, casting a deep shade, and making a palpable diminution of light for several minutes. Myriads of them alighted, and the whole place was covered with them; the main body continuing at a considerable elevation. After gyrating in thick columns for some time over the heads of the troops, the whole living mass moved forward like an immense cloud, and disappeared in the distance. By many they were supposed at the time to be the precursor of the cholera and sickness that almost immediately followed.

Micklethwait; Lieutenants Streatfeild, Caulfeild, Fletcher, Gregory, Thoroton, Hon. H. Handcock, Baillie,* Hoskins, Ingham, Eyre, Wood, and Bradford; Ensign, Cobham; Paymaster Bennett; Adjutant Preston; Quartermaster Walsh; Surgeon Meé; Assistant-Surgeons Thomson and Gibbons. Captain Godfrey Cooper; Ensign McInnis (recently promoted from Sergeant Major); and Assistant-Surgeon Buttler, were left on duty at Varna, in charge of sick men and the baggage, animals and batmen. Captain Greene had been sent home to the depôt from Varna upon promotion, on augmentation of the regiment. Captain Spring had been appointed to the staff as Paymaster and Staff Officer to the Scutari depôt, a duty requiring great perseverance and ability, and which was most efficiently performed by this officer. 1854

The rendezvous was at Baljik Bay, a few miles from Varna; and, on the 7th of September, all got under way, in lines of transports by division. It was a grand scene: the English had about eighty transports, all of large tonnage; the French more, but smaller,—and the whole led by the combined fleet, formed quite a forest of masts. The expedition anchored off Eupatoria on the evening of the 13th, and, moving a few miles south during the night, landed in the Crimea on the 14th of September, at Old Fort near Lake Touzla.

No opposition was offered to the disembarkation of the combined forces. The evening of the 14th proved wet; it rained in torrents all night, and there was not a dry coat in the army; but the next morning was fine, and there was no more rain for some days. The temporary discomfort however, proved beneficial, for the water of Lake Touzla close by was salt and brackish, and the pools of rain-water served the men, till they could be supplied with better from the fleet. Some tents were landed the following day (16th), but were sent on board again on the 18th, together with the men's knapsacks, a light marching order kit, folded in the blanket, only being

* Afterwards Brevet-Major the Hon. R. Baillie Hamilton, he having changed his name.

1854 retained. Officers carried the same on the march, their baggage ponies having been left at Varna; three days, provisions were issued to the whole of the troops.

On the 19th of the same month the army commenced its march, proceeding along the sea-coast to the south, and reached the stream Bulganak towards evening, where, after a slight skirmish between the British cavalry and artillery and those of the enemy, it encamped. Here the first shot was fired between the hostile armies. The march was resumed on the following morning, and a halt was made within a short distance of the river Alma. This river rises in the mountains in the east of the Crimea, and falls into the sea about twelve miles to the north of Sebastopol. The southern bank is formed of almost precipitous hills, intersected by deep ravines. At the mouth of the river the cliffs are several hundred feet in height, and towards the sea they are nearly perpendicular. In the centre of the Russian position was a large hill of cone-like shape, on which extensive batteries and intrenchments had been formed, whilst the crown of the hills was defended by large bodies of infantry. Fatigued though the British were by their long march under a broiling sun, and over the parched Tartar steppes, and accompanied by cholera to the battle-field, all felt most anxious to cope with the enemy. Daybreak saw the allies under arms, but it was noon by the time the Alma was reached.

Both armies advanced on the same alignment, the British with a front of two divisions, in contiguous double columns of companies at half distance from the centre of divisions, with the front and left flank covered by light infantry (Rifle Brigade) and a troop of horse artillery; the second division, under Lieut.-General Sir De Lacy Evans, forming the right of the first line, and touching the left of the third French division under Prince Napoleon, and the light division, under Lieut-General Sir George Brown, the left; the former being supported by the third division, under Lieut-General Sir Richard England, and the latter by the first division, commanded by Lieut-General His Royal Highness the Duke of Cambridge. The fourth division, under Lieut-General Sir George Cathcart, was in reserve. 1854

The light and second divisions having deployed into line advanced to attack the position. The first and third divisions were formed into a line of columns at deploying distance as a second line; and being found to over-lap two regiments of the third division, the 4th and 44th were detached in support of the Brigade of Guards (first of first division), which was immediately opposite to the batteries on on the Kourgane Hill, the principal point of attack. The divisions of the first line having in like manner overlapped, the second division was formed into two lines. The third division was thus in effect thrown into a third line.

This magnificent advance of the British lines under a galling fire, and in face of dense columns of the enemy strongly posted on the heights, and the final capture of the position, are matters of history.

On leaving the third division, the 44th was deployed into line, and, following closely the Grenadier Guards, passed the Alma at and above the bridge at the village of Bourliouk, when the right company, crossing by the bridge, moved up, by order of Sir R. England, and opened fire on a large Russian column which was supporting the work on the hill. The batteries of the enemy having been dismantled, and the position carried shortly afterwards, the regiment was moved to the right, and passing through the leading brigade of the second division proceeded up the gorge or pass in pursuit of a Russian column retreating in all haste and the men throwing away their knapsacks, the British artillery meanwhile playing on the column over the heads of the regiment as it advanced. On gaining the heights above, the regiment was formed in line (with the remainder of the second brigade, third division, which followed closely) and there halted—a pursuit having been decided against by the allied commanders.

The battery of artillery, which was rapidly brought up the hill, formed on the right of the 44th and annoyed the retreating Russian columns, now six or seven hundred

1854 yards distant, with a few well-directed shot and shell. The last shot was fired from this point.

Through mistake the 4th and 44th, as well as the rest of the third division, were omitted from the despatch of the battle; but their places were afterwards recognised by the Commander-in-Chief, on representations of Colonels Cobbe and Spencer; and have been properly* assigned to them in the Government plan of the battle.

Kinglake, in his "History of the Invasion of the Crimea," in the following passage, gives the place of the 44th, without naming the regiment:—

"The whole line, in which the Duke of Cambridge now "moved forward to the attack of the Kourgané Hill, was "more than a mile and a half in length—it was only two "deep; but his right regiment (Grenadier Guards) was "supported by a part of Sir Richard England's division."

This part has already been shown; and the small number of casualties of the 44th—one private killed, and one corporal and six privates wounded—was mainly owing to the judicious deployment of the regiment as soon as detached from the division, by Colonel the Hon. A. A. Spencer, and to his ordering it to lie down in line while under direct fire from the batteries, when not required to advance.

Lieutenant B. S. Hoskins received a contusion, but was not returned as wounded.

The first gun was fired from the fleet, which accompanied the march of the army, at about one p.m., and the last from the artillery of the third division, at 4.15 p.m., so that, in a little more than three hours, the allied armies became masters of a position which the Russian commander had vainly expected to hold for a fortnight.†

---

* With the exception of their position after the action.

† At the Alma the regimental dog "Bob," of the Scots Fusilier Guards, being near the 44th, attached itself to the latter during the battle, having lost his own corps, and remained with the 44th, being fed and taken care of by letter F company for several days during the flank march, and until ground was taken up on the plateau of Sebastopol, when he found his way back to his old friends of the

There is an instance of devotedness connected with this battle which must not be omitted, and which will be always inseparably associated with the 44th Regiment. The following account from Mr. Russell's correspondence, as published in the "Times" of the 24th October, 1854, cannot be improved upon:— 1854

"We have lately been told, again and again, that the "increased civilization and humanity of modern times will "greatly mitigate the horrors of war. To a certain extent "this is undoubtedly true. Our list of Russian prisoners "proves it. Russians easily get rid of such incumbrances "by the summary method of the bayonet. I felt proud of "my country when I saw little groups of English soldiers "tenderly nursing wounded and dying Russians on the "field of Alma, the day after the battle, particularly when "my indignation was hardly calmed at the well-proved "stories of ferocious feeling manifested by these same "wounded. In several instances they treacherously fired "at or stabbed the doctor who had just dressed their "wounds, or the kind Samaritan who had slaked their "parched throats with a draught of water; thus showing "all the inextinguishable hatred of a dying wild beast, "and none of the noble chivalry of a civilized soldier. "Notwithstanding the ingratitude of these poor Russian "slaves, an English doctor, to his eternal honour, volun-"teered to remain behind and endeavour to alleviate the "sufferings of seven hundred wounded Russians, who had "been removed from the field of battle, on the south bank "of the Alma, to the deserted village on its north bank.

"Dr. James Thomson, Assistant-Surgeon of the 44th "Regiment, and his soldier servant, deserve to be held up "as heroes. For four or five days they, and they alone, "had to wait upon and support this enormous mass of "severely wounded men. The task was in many respects

---

Guards. This dog may now be seen preserved in the Royal United Service Institution at Whitehall. He was returned as "missing" at the Alma, and his history is related on the card attached to his stuffed remains; the above circumstance affords the solution of his unaccountable absence.

1854 "a most dangerous one—as we have seen, the patients "themselves were not to be trusted. The Cossacks might "also at any time make prisoners of them on the retreat "of the allied armies. The dead were festering in heaps "around the sick and dying. These two men frequently "had to bury a horrible mass of carcases and fragments "before they could get at some poor wounded wretches. "In this way they must with their own hands have dragged "out and buried some two hundred. There was no food "of any kind for the sick, so the soldier managed to drive "in a stray bullock, and with the aid of some Russian "convalescents (their misfortunes seemed to have "humanised them) he killed it and made some soup for "them. At length Her Majesty's ships, 'Albion,' "'Vesuvius,' and the steam transport 'Avon,' arrived. "The whole crew of the first landed and removed the "wounded on board the 'Avon,' while the 'Vesuvius' "guarded the shore. In the midst of this humane occupa-"tion a Russian force of some four thousand or five "thousand men approached the village, and the sailors "were obliged to hasten on board, as it was beyond the "range of the ship's guns. However, three hundred and "and forty wounded were put on board the 'Avon,' under "charge of their heroic preservers, Dr. Thomson and his "servant. About forty were left behind, and many of "these poor fellows who had previously seemed unable to "walk, endeavoured with all their might to hobble after "their more fortunate comrades. The next day the "'Avon' proceeded with them to Odessa. Now, surely, "when the Humane Society rewards a man who saves one "single individual, society will not fail to do something "for two men, who, under such dreadful trials, saved the "lives of three hundred and forty."

The sequel remains to be told. Dr. Thomson having attended his charge to Odessa, where they were given over to the Russians, returned to the Crimea and landed at Balaklava *en route* to rejoin; his one suit of uniform bearing unmistakeable tokens of his labours in the cause of humanity. He never reached his regiment. Fatigued and weakened by his benevolent exertions, he fell a ready victim

to the cholera, which was raging at Balaklava, at which 1854
place he died, on the 5th of October, the morning after his arrival.*

The name of the soldier who accompanied him in his humane mission is private John Margrath 44th Regiment. He served for many years after in the Regiment, and bore an excellent character.

During the 21st and 22nd the armies halted on the Alma heights, and were engaged in burying the dead of both sides, in carrying the British and French wounded to the fleet and the Russian wounded to the houses in the village below. The march was resumed on the 23rd of September, to the high ground beyond the River Katcha, where the Russians offered no opposition—having, it was ascertained, with scarcely a halt, retreated to their fortress of Sebastopol. On the 24th the troops moved to beyond the river Balbec, meeting with no resistance; and on the following day, by a flank march, towards the south of Sebastopol. The army came in collision on this day, at McKenzie's Farm, with a Russian rear-guard, or a large provision and baggage escort, moving from the city.

Marching thence, in the evening, after dark, the third division arrived at and bivouacked on the banks of the Tchernaya, at two o'clock in the morning of the 26th of September. Again moving, the army arrived the same afternoon at Balaklava. Possession was at once taken of this harbour, and the place put in a state of defence by the British. On the 27th, the third and fourth divisions marched to the neighbourhood of Sebastopol, halting on ground afterwards occupied by the French left attack. After several changes of position, the division at length on the 2nd of October, settled down to its fixed position

* An obelisk, with an inscription commemorative of his heroism, has been erected by subscription to the late Assistant-Surgeon Thomson near his native town, Cromarty; and a foundation for the encouragement of learning established there, called "the Thomson Bursary."

1854 throughout the siege, on the left of the British attack.* A small proportion of tents was sent up on the three following days for the regiment.

Every nerve was now strained to break ground before Sebastopol. On the 28th the Royal Artillery commenced disembarking, and in five days the landing of the British siege train *matériel* was effected. It was a busy time as well for the assailants as for the enemy; for whilst the latter made determined efforts to frustrate those of the besieging troops, the former were fully occupied in conveying from Balaklava —seven miles distant from the camp—guns, ammunition, military stores, and provisions. Ground was broken and trenches opened on the 10th of October.

The first bombardment of Sebastopol took place on the 17th of October, and lasted until the evening of the 19th, when it became evident that the place could not be assaulted so early as was at first anticipated. Several changes in the works of attack, therefore, became necessary; the principal one of which it was only possible, owing to the Russian fire, to effect under cover of night. The regiment was in the left attack (Greenhill) on the 17th, and had three men killed, and Lieutenant Wood, one sergeant and four rank and file wounded. Whilst in the trenches, on the 20th of October, Captain Andrew Browne,† Lieutenant Michael Bradford, and Assistant-Surgeon John Gibbons, were wounded, and one drummer killed, and six rank and file wounded by the explosion of one shell. On the following

---

* A detachment of sickly men had been left at Balaklava, under Captain Howard and Lieutenant Richard Fitz-Richard Eyre. The latter died of cholera and dysentery, on the 15th of October. This detachment was present at the action of Balaklava.

† His wound occasioned the amputation of the right arm and part of the left hand. Assistant-Surgeon Gibbons distinguished himself on this occasion, having dressed the wounds of all who were hurt before looking to or even mentioning his own, which although slight was painful, a small piece of shell having become embedded in his leg. Lieutenant Bradford retired from the service the following year, and died shortly afterwards from the effect of his wound.

morning Sergeant-Major O'Neill was taken prisoner 1854
whilst withdrawing the advanced sentries. He remained in the hands of the Russians for twelve months.

On the 25th of October occurred the famous cavalry action at Balaklava, and on the following day the sortie of the enemy against the right of the British position. The regiment was not present at either of these actions, being on duty in the trenches. Strengthened by numerous reinforcements, the Russian commander—Prince Menschikoff—determined again to attack the allied armies, and the result was the sanguinary battle of Inkerman on the 5th of November.

On the return from the trenches of a party of the 44th, on the morning of the 5th of November—the greater part of the regiment having formed a portion of the relief—a rapid roll of musketry, accompanied by the deeper roar of artillery, was heard from the right of the British position. The Russians—under cover of a thick fog and a drizzling rain—had established themselves during the night on the heights above the Inkerman Valley, within a few hundred yards of the picquets of the second division. As day dawned, they advanced, and endeavoured to turn the flank of the British position, of which the second division held the extreme right. A brisk cannonade, which riddled their camp, quickly roused the regiments of the second division, who, flying to arms, were soon hurried to the support of the picquets then doing their best to check the advance of the heavy Russian columns. Thus commenced that famous "soldiers' battle," where the men of the different regiments, in many instances, mixed together, skirmished, or formed themselves into parties under the nearest officer, and so, in a measure, creating order out of confusion, disputed the ground, inch by inch, till the arrival of the light division and brigade of Guards infused some regularity into the array. The Artillery now hurried up from all parts of the camp, and the fourth division, under Lieut.-General the Honourable Sir George Cathcart, also coming up formed in support. This gallant officer was killed in an attempt to turn the Russian left. The

1854 third division—a great proportion of which was in the trenches, under Brigadier-General Eyre—was held in reserve, occupying ground in advance of what was afterwards called Cathcart's Hill, and detaching its two strongest regiments—the 1st and 50th—under Brigadier-General Sir John Campbell, in support of the left of the light division. A division of the French, detached in aid of the British, arrived on the field soon after. It not being the object, however, of this record to enter into a minute description of an action in which the 44th was not actively engaged, suffice it to say, that in the afternoon, after the most obstinately contested battle of the war, and one on which depended the maintenance of the siege, and perhaps the very existence of the allied armies, the Russian columns were compelled to retreat, with tremendous loss. It appeared that, by a mistake in the darkness and fog of the night, the Russians had concentrated on the Inkerman heights, instead of attacking the British position at two points at once, as had been intended. Their columns thus crowded together were unable to deploy, and the Minié bullets of the British troops penetrated them with telling effect.

The greater part of the 44th was, as has been shown, engaged in defending the trenches, on which the Russian batteries kept up a heavy cannonade; and the men off duty—one hundred and fifty in number—under Major Feilden, with the reserve, were in advance at Cathcart's Hill.

Lord Raglan expressed his sense of the services of the third division in the following terms:—

"I must likewise express my obligations to Lieut.-
"General Sir Richard England, for the excellent dis-
"position he made of his division, and the assistance he
"rendered to the left of the light division, where
"Brigadier-General Sir John Campbell was judiciously
"placed, and effectively supported Major-Gen. Codrington;
"and I have great pleasure in stating that Brigadier-
"General Eyre was employed in the important duty of
"guarding the trenches from any assault from the town."

His Lordship, in a subsequent despatch, bore testimony 1854
to the share taken by the division in all the siege operations, as will be seen by the following extract:—

"The third division was only partially engaged; but "having been actively employed in all the siege operations, "Lieut.-General Sir Richard England avails himself of "the opportunity to mention, in terms of high approbation, "the staff officers and officers in command of regiments." The name of Colonel the Honourable Augustus Spencer, commanding the 44th, was specially mentioned.

As a mark of Her Majesty's recognition of the meritorious services of the non-commissioned officers in the Crimea, the Queen directed that a sergeant* should be selected from each regiment of cavalry, guards, and infantry, of the army under Lord Raglan, for promotion to a commission, the date to be that of the battle of Inkerman. The thanks also of both Houses of Parliament was voted for this victory and that of the Alma.

In December, 1854, the regiment was augmented to sixteen companies; eight companies, consisting of one thousand rank and file, before Sebastopol, four as a reserve at Malta, and four as a depôt. The entire strength was directed to be one hundred and nine sergeants, forty-one drummers, and two thousand rank and file.

This augmentation occasioned an additional lieut.-colonel to be added to the regiment, and Major Charles William Dunbar Staveley was promoted to that rank, his commission bearing date 12th of December. Major Robert Feilden became the senior major and Captain Patrick William MacMahon was promoted to the vacant majority.

---

* The following non-commissioned officers of the 44th were appointed to commissions in the regiment whilst serving in the Crimea: — Quartermaster Sergeant Richard Gilham Thomsett, appointed quartermaster on 27th July, 1855, and paymaster on 25th January, 1856 (died 15th December, 1861); Sergeant-major Francis O'Neill, promoted ensign on 6th November, 1855, and adjutant on 1st February, 1856; Sergeant-major William Hart, promoted quartermaster on 25th January, 1856; Colour-Sergeant J. Rowe received a cornetcy in the Land Transport Corps.

1854 Christmas,* before Sebastopol, might be compared to that passed by the 44th in Affghanistan ; in both instances the customary greetings of that festive season must have seemed almost a mockery. The weather, which, with the exception of the night following the landing at Old Fort, had been particularly fine, indeed almost cloudless, broke at the beginning of November, and on the 14th of that month the Black Sea and the Crimea were visited by a tremendous gale, accompanied by driving rain and sleet, which left not a tent standing in the allied camp, and wrecked numbers of vessels both within and outside the harbour of Balaklava; amongst others a fine steamer, the "Prince," with a most valuable cargo, and winter clothing for the whole army, was totally lost. There were also lost on board this vessel private stores for the 44th, consisting of two thousand flannel shirts, one thousand pairs of socks, nine hundred pairs of cloth trousers, and five hundred shell-jackets. This loss it was at first impossible to remedy, but after a while curious sheepskin and rabbitskin coats were served out, but not in sufficient numbers for all demands. From this storm may be dated the sufferings and privations of the winter siege. Cholera, from which the army had never been free since first attacked at Varna, increased daily: scurvy and frost-bites sent numbers to hospital: and during November, and the four following months,† the

---

* On Christmas Day, 1854, the 44th had short rations issued to it, and at so late an hour that the men for duty had no time to cook their morsel before going into the trenches. On the following Christmas Day there was a gathering of twenty-two officers, and a happy evening was spent; song and merriment prevailed, and although cold and snow were without, warmth was within—the many incidents and memories of the war giving a zest to the festivity.

† Deaths, 44th Regiment:—

During November, 1854	32
,, December, ,,	36
,, January, 1855	57
,, February ,,	111
,, March ,,	77
	313

The total casualties sustained by the 44th from disease during

deaths in the 44th amounted to three hundred and thirteen non-commissioned officers and men. The weather increasing in severity as the winter advanced, the trench duties, without shelter from snow or rain, became most arduous; provisions were besides scarce in the front, owing to the failure of the transport, and the impassable condition of the roads. This state of things being aggravated by the fact that both officers and men were frequently on duty two and even three nights in succession, many a hardy soldier succumbed to this combination of privation and fatigue.* 1854

Towards the end of December a draft for the regiment arrived at Balaklava, and in January it joined the headquarters at the front. The new year brought with it no improvement; on the contrary, the commissariat transport having broken down, fatigue parties had to be sent to carry up warm clothing and rations from Balaklava. Men frequently died in the trenches from cold and exhaustion; at length a railroad was commenced from the above place to the camp, which proved most serviceable, and towards the middle of January, 1855, a favourable 1855
change in the weather became manifest; the health of the regiment however as yet showed no improvement.

Lieut.-General Sir Frederick Ashworth was appointed Colonel of the 44th on the 8th of February, 1855, the

---

the service of the regiment in the Crimea, amounted to fifteen sergeants, nine drummers, and three hundred and seventy-seven rank and file. From first to last the 44th were in the *front* on the plateau of the left attack before Sebastopol.

* The band of the regiment suffered equally with their comrades; they had at first been told off as stretcher-carriers, and owing to the frequent calls on their services on the march from Old Fort, principally from cases of cholera, their instruments became a secondary consideration, and were unavoidably much damaged. As the pressure for duty-men became greater the bandsmen were sent to the ranks, and by the end of the winter there were left neither bandsmen, instruments, nor music. This, it will be seen, was the third time within thirty years that the band had been destroyed—viz., on the Ganges in 1823 in Affghanistan in 1842, and in the Crimea in 1855.

1855 decease of General the Honourable Sir Patrick Stuart, G.C.M.G., having occurred on the 7th of that month.

Early in February the Regiment received much praise from Major-General Eyre, at his inspection, for its cleanly and efficient state, despite the difficulties to be contended with. The light company (Captain Faussett's) in particular was quoted as a pattern for other regiments of the brigade. This officer was shortly afterwards appointed brigade-major by General Eyre.

This month proved the worst which had been experienced by the regiment. On the 1st of March the strength in non-commissioned officers and men was nine hundred and thirty-five, but of this number four hundred and ninety-five—more than one-half—were reported "sick." One hundred and eleven were struck off as dead since the 31st of January, 1855.

During March the operations of the siege were carried on with greater activity, and the second bombardment of Sebastopol was opened on the 9th of April, but terminated in eight days, without any apparent result. During May the health of the regiment improved, and in June it was reported as "good."

A third bombardment commenced on the 6th of June, in the afternoon instead of at daybreak, as in the two former instances. This also ended in a few days. By the 16th of June new batteries had been completed, which, it was anticipated, would place the besiegers in a position to resume the offensive with the utmost vigour. Accordingly, on Sunday the 17th of June, the fourth bombardment was commenced, and an attack on the following day determined upon. This afforded the regiment, which on the previous day had been armed with Enfield in lieu of Minié rifles, an opportunity of distinguishing itself in the attack and occupation of the Cemetery at the head of the Dockyard Creek.

It was arranged that the French assault on the Malakoff should take place at three o'clock in the morning of the 18th, and before that hour the British Commander and head-quarter staff, with other officers, arrived at the appointed post. As day broke the French commenced their operations from the Inkerman attack, and, as their columns issued from the works, they encountered the most serious opposition both from musketry and the guns in the works, which had been silenced the previous evening. Observing this, Lord Raglan forthwith ordered the British columns to move out of the trenches upon the Redan. 1855

By half-past three it was perceived that General Pelissier had not succeeded in his attack upon the Malakoff. All the allied batteries were therefore ordered to resume their fire as heavily as possible. However, at about half-past seven, the firing slackened in consequence of the attack being relinquished.

Lieutenant-General Sir Richard England had been ordered, whilst the direct attack upon the Redan was in progress, and as a co-operative movement from the left of the English position, to send one of the brigades of his division, under the command of Major-General Barnard, down the Woronzoff Ravine, with a view to afford support to the attacking columns on his right; and the other brigade, under Major-General Eyre, was, still further to the left, to threaten the works at the head of the Dockyard Creek. Major-General Barnard, proceeding down the Woronzoff Road, was placed in position on the right of the ravine, ready to co-operate with the columns of attack on the right, whilst Major-General Eyre, with the second brigade, moved down by the "Valley of Death" to another ravine which separated the left of the British from the right of the French advanced works, for the purpose of attacking the enemy's ambuscades, and making a demonstration on the head of the Dockyard Creek.

The brigade under Major-General Eyre, consisting of the 9th, 18th, 28th, 38th* and 44th Regiments—total strength about two thousand bayonets—moved off on this service between one and two o'clock a.m., and, in attacking the

* The 38th, from the first brigade, third division, had recently been exchanged with the 4th Regiment.

1855 first of the ambuscades, the troops were anticipated by the French, who cleverly took them on their left flank as the British advanced in front, and made several prisoners. Beyond this the French had no orders to co-operate, and Major-General Eyre therefore pushed on an advanced guard under Major Robert Feilden, of the 44th, composed of marksmen from each corps, formed into two companies under Captain Robinson and Lieutenant the Hon. H. Handcock,* of the 44th, and a captain and subaltern of the 28th Regiment, supporting it on the right by the 44th and 38th, and on the left by the 18th, and keeping at first the 9th and 28th in reserve. The Russian picquets were at once driven in by the advance, and the ground to be occupied by the brigade cleared.

No estimate could be formed of the strength of the enemy, who occupied a strong position, their right resting on a mamelon and their left on a cemetery. Marksmen in rifle pits occupied all these points, and the intervening ground was intersected and the road barricaded with stone walls, which the men were obliged to pull down, under fire, before they could advance. In rear of this position, towards the fortress, several houses were occupied by the Russians, and there were reserves of them seen in the rear. This position, under the fire of the guns of the fortress, was strong, and it could not be expected to be carried and retained without incurring a considerable loss, which was experienced both in officers and men, who all nobly discharged their duty. The 18th pushed on and occupied some houses immediately under the garden-wall battery, on the left; and the 44th, swarming into the advanced houses on the right, kept up a continuous fire on the embrasures at the head of the Dockyard Creek. After taking possession of some houses in front, the 38th endeavoured to turn the flank of a battery which annoyed the troops in front. These parties were subsequently

* Lieut Handcock was promoted to a company in the regiment in succession to one of the death vacancies of the following day. He was afterwards killed by a tiger, while with the regiment in India, on the 16th of December, 1858.

reinforced from time to time by the 9th Regiment, the 28th being drawn up in line in rear to support the whole. 1855

Having driven the Russians from all these points, the brigade continued to occupy them, with a view to ulterior movements, in case of the attack on the right proving successful, and until it should be decided what portion of the ground it was considered advisable to retain for siege operations.

It had been expected that the anniversary of Waterloo would have proved more fortunate, but as the attack was not successful, and as the town front was not attacked, Lord Raglan gave orders for the withdrawal of Major-General Eyre's brigade, the ground won being too far in advance to be permanently occupied for siege operations; the high ground in the Cemetery however was retained, and the advanced parallel of the left attack was prolonged to it on the following night. The whole position was, nevertheless, held until dark, the troops having been exposed throughout the day to a concentrated fire from the guns of the fortress, in addition to some field guns expressly brought to bear upon them.

At dusk the fire slackened, and after removing the wounded to the rear, the brigade was gradually withdrawn, the command of it having devolved on Lieut.-Colonel Adams of the 28th, in consequence of Major-General Eyre having received a wound in the head in the early part of the day, which incapacitated him at the close of the action from attending to his duties. No attempt was made to molest the troops on their retiring; their casualties were considerable, amounting to thirty-one officers, forty-four sergeants, and four hundred and eighty-seven rank and file, killed or wounded.

The 18th, 38th, and 44th were the most advanced on this day. It was ten minutes past three when they went into action. They got (for the first time for any part of the British army) actually into the town of Sebastopol, being so close to the Creek and Garden Bat-

1855 teries that the Russians could not depress their guns sufficiently to oppose them. The buildings and gardens of this portion of the town, the suburb at the head of the creek—the inhabitants having only just left them—were occupied by these regiments; but all were enfiladed and exposed to a plunging fire from the Great Redan and Barrack Batteries. Five hundred and sixty-two officers and men—more than a quarter of the brigade—were the casualties of this day, and the conflict might, in truth, be named the "Infantry Balaklava." The brigade was altogether under fire about eighteen hours, having remained in the position until nightfall, when it slowly retired at about half-past nine, under cover of darkness, and reached the camp between ten and eleven p.m., having been absent just twenty-one hours. Soon after its return to camp there was an alarm, when the fatigued remnant of the brigade again fell in; this proved to be a false alarm. The affair in the Cemetery was, according to Lord Raglan's despatch, most daring, and so hot was the fire from the Garden, Barrack, and Creek Batteries, that many of the wounded officers and men had to be left in the open all day; at dark, and before the brigade was withdrawn, all were brought in that could be found; the dead were left where they fell, and although search was made for them after dark were, with a few exceptions, never recovered or buried. The following night the new trench was opened in the Cemetery, when the Russians shelling the working-parties unceasingly, and setting fire to the long grass with carcases and fire-balls, the effect, more brilliant than welcome to the soldiers at work—their exact position being thus discovered—was heightened by a bombardment of the town and sea batteries by the allied fleet.

In his despatch (of which a copy was forwarded by Lord Raglan to Lord Panmure), Major-General Eyre stated that he could not "sufficiently express his sense "of the conduct of the officers, non-commissioned officers "and men on this occasion. The conduct of all was so "exemplary, during this trying day, that he could scarcely with justice particularize individuals."

Colonel the Honourable Augustus Spencer, commanding the 44th Regiment, who was wounded, and Lieut.-Colonel Staveley, who succeeded to the command on the former being obliged to quit the field, were specially thanked for their assistance, together with Major Feilden, who commanded the advanced guard. The valuable services of Brigade Major Captain Faussett and Assistant-Surgeon John Gibbons of the 44th, were also acknowledged,—the latter, in conjunction with Assistant-Surgeon Jeeves of the 38th, whilst exposed to a most galling fire, having exerted himself in the field in attending to the wounded in so zealous and humane a manner as to call forth special notice. Colour-sergeant James Donelan and Private Robert Thimbleby were particularly named by Major-General Eyre for gallantry—the latter for assisting wounded comrades while exposed to a heavy fire. 1855

The 44th had the following officers killed or wounded on the 18th of June:—

Captains Bowes Fenwick (died following day) Honourable Charles Welbore Herbert Agar (died same evening), William Henry Mansfield (died of wounds, 28th June), and Francis William Thomas Caulfeild (died following day, 19th); Colonel the Honourable Augustus Almeric Spencer, and Lieutenants Joseph Logan and T. Orton Howorth wounded; Lieutenant B. S. Hoskins,* although not so reported, was also wounded; one drummer and sixteen rank and file killed; eleven sergeants, two drummers, and ninety-six rank and file wounded; seven men were missing, supposed to have been killed, but, owing to the roughness of the ground, were not found by

---

* Lieutenant Hoskins was promoted to a company in the regiment in August of this year, but retired from the service in 1858. He joined the English Volunteer Legion of Garibaldi's army in 1860, with which he served as major of the brigade. He subsequently took service as captain in the army of the Confederate States of America, and was killed in action in 1863.

1855 the party sent after dark to bring in the killed and wounded.*

Of six captains who went into action four were killed, and buried in front of the camp of the third division. A monument was placed over them, and their names also inscribed on the general regimental monument afterwards erected on Cathcart's Hill. Great credit was gained by the regiment for its steadiness and endurance on this trying day.

On the receipt of the news in England, a telegraphic despatch was sent to the Crimea, and being published to the army, afforded much gratification to the troops. It expressed Her Majesty's grief "that so much bravery should not have been rewarded with merited success." An additional interest attaches to the order issued on the occasion, from its being the last emanating from Lord Raglan, who died about nine o'clock on the 28th of June, the date of the order. This was a sad blow to the army, for his Lordship was esteemed by all; "and his country," to use the Minister of War's expression, "had, indeed, "been deprived of a brave and accomplished soldier, a "true and devoted patriot, and an honourable and dis-"interested subject."

Lieut.-General Sir George Brown being absent, on account of ill-health, Major-General Simpson, the chief of the staff, as next senior officer, assumed the command; the appointment of Commander-in-Chief in the Crimea, with the local rank of general, being subsequently conferred upon him by the Queen.

Colonel the Honourable Augustus Spencer was appointed, on the 30th of June, to the command of the first brigade of the fourth division. He appointed Captain Robinson as his Aide-de-Camp. Lieutenant-Colonel

* The number of men of all ranks "under arms" of the 44th, on the 18th June, 1855, was four hundred and sixty-six; the casualties, therefore, one hundred and thirty-three in number, amounted to considerably more than one-fourth.

Staveley succeeded to the command of the regiment. 1855

Captain Levett Thoroton, of the 44th, was slightly wounded in the trenches on the 28th of July. This month had passed in repelling occasional sorties from the enemy, and in strengthening and improving the advanced works; which, by the 10th of August, had become so close to the place, that any further approach could not be made without great difficulty.

On the 2nd of August, Lieut.-General Sir William Eyre (who had been made a K.C.B.) succeeded to the command of the third division, and Colonel Trollope, C.B., of the 62nd Regiment, was appointed to the second brigade.

Fresh drafts arrived for the 44th from Malta and England towards the end of this month.

The hope of the Czar that he would be able to relieve the garrison was at length dispelled by the battle of the Tchernaya, gained on the 16th of August by the French and Sardinians. The time had at length arrived for the final bombardment of Sebastopol, and it was arranged that the firing should commence steadily on the morning of the 5th of September, and be increased as the day progressed, care being taken to reserve a sufficient quantity of ammunition for a very heavy fire just previous to the assault, ordered to take place on the morning of the 8th. A tremendous cannonade was commenced by the French about five o'clock a.m., the British batteries opening at the same time on the Redan and Malakoff. The bombardment continued during the 6th and 7th, and on the morning of the 8th the whole of the batteries were in full play. At a few minutes before noon the French signal was given, and the Malakoff gained without loss; this work was retained by the French, who defeated every attack made to regain it.

After experiencing very heavy loss, and the Russians bringing up their reserves, the British proved unable to maintain themselves in the salient angle of the Redan,

1855 which they had penetrated, and orders were given to withdraw. The British share in the assault was entrusted to the second and light divisions, under Lieut.-General Sir William Codrington.

The attack was to have been renewed on the following morning, with the Highlanders, to be supported by the third division, General Simpson having made arrangements for this object with Lieut.-General Sir Colin Campbell, commanding the Highland division, and Major-General Sir William Eyre, commanding the third division; the advanced trenches were accordingly occupied by the Highland division during the night of the 8th of September; however, about eleven o'clock the enemy commenced exploding his magazines, and Sir Colin Campbell, having ordered a small party to advance cautiously and examine the Redan, found that the work had been abandoned. It was not, however, considered necessary to occupy the place until daylight.

Extensive explosions were heard during the night, and before daybreak the whole city was in a blaze. It soon became known that the Russians were withdrawing from the south to the north side, by means of the bridge recently constructed across the harbour. This was subsequently disconnected and conveyed to the other side, and the ships of war all sunk during the night.

In this manner ended the great siege of Sebastopol, and the termination of their labours was hailed with satisfaction by all ranks. The following officers of the regiment were present at the fall of the place:—Lieut.-Colonel Staveley, C.B.; Major MacMahon; Captains Greene, Fletcher, Thoroton, Baillie, Preston, Gregory, Honourable Henry Handcock, and Hoskins; Lieutenants Ingham, Cobham, Smith, Pigott, Bower, Wood (adjutant), Walters, Kendall and Rogers; Quartermaster Thomsett, Surgeon Mee, Assistant-Surgeons Gibbons, Butler and Johnson; Colonel the Honourable A. Spencer, Brigadier-General; Captain Robinson, aide-de-camp; and Brevet Major Faussett, brigade-major.

Throughout the siege the regiment had been conspicuous for the vigilance of its sentries, and the soldier-like performance of its trench duties. It was proverbial in the division that everything was safe when the 44th were in the advance, and singularly enough, a sortie was never attempted whilst the regiment was in the advanced trenches. 1855

On the morning of the 9th of September, the alarm bell of the Redan, now in possession of the regiment, was brought away as a trophy.

No further opportunity was afforded for the 44th to distinguish itself; but its commanding officer, Brigadier-General the Honourable Augustus Spencer, was sent with a force, comprising the 17th, 20th, 57th, and 63rd Regiments, two battalions of Royal Marines, and a brigade of French, for the capture of Kinburn (situated on a tongue of land commanding the mouth of the Dnieper), which was effected in the following October. Captain Baillie, of the regiment, accompanied this expedition as extra aide-de-camp. On the 26th of October Sergeant-Major O'Neill returned from his captivity in Russia.

Decorations, medals and rewards, were bestowed upon the army in the Crimea. The words "ALMA," "INKERMAN," and "SEVASTOPOL," were authorised by royal authority to be emblazoned on the regimental colour of the 44th, in commemoration of its services. Colonel the Honourable Augustus Spencer, who had received the local rank of Brigadier-General in Turkey, was made a Companion of the Order of the Bath, and received the distinguished service pension. The following officers of the regiment received a step in rank, together with the decorations, &c., hereafter enumerated:—

Major Charles William Dunbar Staveley was promoted Lieut.-Colonel, and was made a Companion of the Bath. Majors Patrick William MacMahon and Andrew Browne were promoted to the brevet rank of Lieut.-Colonel. Major Robert Feilden received the brevet of Lieut.-Colonel,

1855 and Captains William Faussett, John Robinson, Richard Preston, and William Fletcher, that of Major.

The Imperial Order of the 4th Class of the Legion of Honour was conferred on Colonel Spencer; and Brevet Lieut.-Colonels MacMahon and Browne, Brevet Majors John Robinson and Richard Preston, Assistant-Surgeon Gibbons, and Private Robert Thimbleby, were awarded the insignia of the 5th Class (Knights) of the same Order, the latter for his distinguished conduct in the Cemetery.

Colonel Spencer likewise received the 3rd Class of the Turkish Order of the Medjidie, and Lieut.-Colonels Staveley, Feilden, MacMahon, Browne; Brevet Majors Faussett, Robinson, Preston, William Fletcher; Captains Frederick William Gregory, Levett Thoroton, the Honourable Henry Handcock, and Lieutenant Alexander William Cobham, received the 5th Class of this Order.

In addition to the medal authorised by the Queen, the Emperor of the French, the King of Sardinia, and the Sultan also gave medals. A selection, however, was made in awarding the Sardinian and French war medals, which were not issued to all.

The Sardinian medal was conferred on Colonel the Honourable Augustus Spencer, C.B.; Lieut.-Colonel Staveley, C.B.; Brevet Lieut.-Colonel MacMahon, Brevet Majors Faussett and Fletcher, and Captain Robert Baillie; Lieutenant and Adjutant William Arthur Wood, Privates William Doole and William Woodgate.

The French war medal was conferred upon Quartermaster-Sergeant Denis Reddin, Colour-Sergeant James Donelan, Sergeant Thomas Brown; Corporals John Drenon and Robert Murray, Lance-Corporal Denis Canty, Privates James Edlow, John Burnside, and Thomas McCarthy.

British medals for "Distinguished conduct in the field" were conferred on the following non-commissioned officers and men:—With annuity of £20, on Hospital Sergeant Thomas Austin;* with gratuities of £15, £10,

* On the death of this non-commissioned officer, in 1860, Quartermaster Sergeant Denis Reddin received the medal and a similar gratuity for his good service in the Crimea, and afterwards in China.

and £5, according to rank, on Sergeant-Major William Hart, Corporals Jerome Cunningham, Goen Powell, Patrick Torpy, and William McWhiney; Privates Henry Carroll, Thomas Crawford, Robert Crookshanks, Thomas Eade, Alfred Fawkes, John M'Gann, John Magrath, John Robinson, James Samples, and George Saunders. 1855

Sergeant William McWhiney, afterwards promoted, received on the 17th June, 1857, from the hands of Her Majesty, for repeated acts of devotion and gallantry, the Victoria Cross.*

In February, 1856, a suspension of hostilities took place, and this was followed by a Treaty of Peace, which was signed at Paris on the 29th of March, and proclaimed to the Army on the 2nd of April. Preparations were shortly afterwards made for the break up of the Army, and on the 25th of June the regiment embarked at Balaklava on board Her Majesty's ship "Colossus," and arrived at Spithead on the 18th of the following month, when the 44th and other regiments on board ship were visited by the Queen in her steam yacht. On landing at Portsmouth, on the 19th of July, it proceeded to Aldershot, where it was inspected by Her Majesty; and on the 4th of August moved to Dover. The regiment, 1856

* "Sergeant William McWhiney (as appeared in the *London* "*Gazette*) volunteered as sharp-shooter at the commencement of the "siege of Sebastopol, and was in charge of the party of the 44th; "was always vigilant and active, and signalized himself on the 26th "of October, 1854, when one of his party, Private John Keane, 44th "Regiment, was dangerously wounded in the Woronzoff Road, "at the time the sharp-shooters were repulsed from the quarries by "overwhelming numbers. Sergeant McWhiney, on his return, took "the wounded man on his back and brought him to a place of "safety. This was under a very heavy fire. He was also the "means of saving the life of Corporal John Courtenay. This man "was one of the sharp-shooters and was severely wounded in the "head on the 5th of December, 1854. Sergeant McWhiney brought "him from under fire, and dug up a slight cover with his bayonet, "where the two remained until dark, when they retired. Sergeant "McWhiney volunteered for the advance guard of Major-General "Eyre's brigade in the Cemetery, on the 18th of June, 1855, and "was never absent from duty during the war."

1856 which on return from the Crimea mustered seven hundred and ninety-eight men of all ranks exclusive of officers, marched to the camp at Shorncliffe on the 25th of September, under the command of Lieut.-Colonel Staveley, C.B.; Colonel the Honourable A. A. Spencer having been appointed to a brigade at Aldershot, and placed on half-pay.*

In November, 1856, the regiment was reduced by four companies (Malta reserve); the head-quarters remaining at eight and the depôt at four companies: the latter continued at Walmer.

The 44th was not destined to remain any lengthened period on home service. In consequence of the notorious Indian Mutiny, reinforcements had to be sent at a moment's notice to the several Presidencies; the regiment
1857 proceeded from Shorncliffe to Portsmouth on the 20th of July, 1857, and in a few days received orders to embark for Madras. Two companies from the depôt joined head-quarters at Portsmouth, where the regiment remained whilst tonnage was being taken up.

On the 18th of August new colours were presented to the 44th by Major-General the Honourable Sir James Yorke Scarlett (afterwards Adjutant-General), who then commanded the troops in the south-western district. After the customary military ceremonies had been carried out, and the new colours consecrated by the Rev. E. W. Milner, M.A., Garrison Chaplain of Portsmouth, Sir James delivered them to the two majors, Lieutenant-Colonels MacMahon and Browne, who carried the old Crimean colours.† In a most soldierlike speech, the gallant general recalled to the memory of all present the several campaigns through which the 44th had served.

---

* Colonel, afterwards Major-General, Spencer subsequently served in command of a division in the Madras Presidency.

† The Crimean colours were afterwards placed in St. Peter's Church, Colchester, the county town of the 44th (East Essex) Regiment, where a suitable tablet has been erected to the memory of the officers and men who fell in the Crimea.

He showed that the colours of a regiment mark its identity, 1857
and that those of the 44th, emblazoned as they were with the names of the memorable victories gained in Egypt, at Badajoz, Salamanca, and other hard-fought battles and sieges in the Peninsula, Bladensburg, Waterloo, and, in later times, Ava, Alma, Inkerman and Sebastopol, were such as no corps had more reason to be proud of. Pursuing this theme, in thrilling words Sir James remarked, that this was "no featherbed regiment;" that it had fought for king and country in all parts of the world, in the most brilliant successes, as also in the direst disasters; and that the names on its colours called up recollections which redounded to its honour and credit. Dwelling specially on the gallantry shown in the preservation of the colours at Quatre-Bras and in Affghanistan, and paying a well-merited tribute to the unparalleled devotion at the Alma of Dr. Thomson and Private John Magrath, Sir James, after expatiating on the steadiness and soldierly bearing of the regiment in the attack on the Cemetery before Sebastopol, expressed a wish that as the 44th were about to embark for India, success and victory might still be enchained to their standards, for he felt certain that under no circumstances would they fail to defend them. After a touching allusion to the officers who had fallen in the Crimea, and to Lieutenant Souter, who preserved the colour at Cabul, Sir James concluded a spirited and martial address (the mantle of his father, the celebrated judge, orator, and advocate, seeming to have falling on him) by fervently hoping that the 44th Regiment would preserve the honours of the old colours, and reap new distinctions under those now entrusted to their safe keeping.

Three companies, under Lieut.-Colonel MacMahon (promoted to the second lieutenant-colonelcy on the augmentation of the regiment to the Indian Establishment), embarked on the 26th of August, and seven companies and head-quarters, under Colonel Staveley, C.B., followed on the 29th of that month. The depôt was shortly afterwards moved from Walmer to Colchester.

1857 The women and children of the regiment were left at home.

The head-quarters having proceeded in the steamer "Khersonese," arrived at Madras on the 28th of November; but the three companies which first embarked, being in a sailing vessel, did not arrive at their destination until the 12th of January, 1858.

1858 On the 2nd of August, 1858, Major-General Thomas Reed, C.B., was appointed to the colonelcy of the 44th Regiment, in succession to Lieut.-General Sir Frederick Ashworth, deceased.

1859 During the years 1858 and 1859 the regiment remained at Fort St. George, Madras, whence it proceeded on active service, for which it had been specially selected on the recommendation of Lieut.-General Sir Patrick Grant, K.C.B., Commander-in-Chief in the Madras Presidency, who addressed the regiment in most flattering terms on its departure from India.

1860 The war with China in 1860 was owing to the Emperor having refused to ratify the treaty which had been signed two years previously by his ministers at Tien-Tsin. In this campaign, as in the Crimea, the French and British forces acted again together. The English troops were commanded by Lieutenant-General Sir James Hope Grant, K.C.B., and the 44th was one of the regiments selected for this service.

Five companies of the regiment embarked at Madras on the 31st of January for China, under the command of Lieut.-Colonel MacMahon; and the head-quarters, under Colonel Staveley, followed on the 3rd of March. The strength on embarkation consisted of ten companies, of thirty-five officers, and one thousand one hundred and seventy-six men of all ranks. The women and children of the regiment arrived at Madras, from England, after the first detachment had embarked, and were again left behind. After landing at Kowloon, on the mainland opposite Hong Kong, the 44th proceeded on the 15th of May to the north of China. Colonel Staveley having been appointed to command the first brigade of the first division in this expedition, the command of the regiment devolved upon Lieut.-Colonel MacMahon. Captain Hon. R. Bailie Hamilton was appointed major of brigade, and Ensign Irvine orderly officer to Brigadier Staveley. The regiment furnished detachments in charge of parties of the Coolie corps proceeding in seven transports, and to which Lieutenant Howorth was attached. Several picked men of the regiment were also permanently employed with it as orderlies. Lieutenant Acklom and Ensign O'Neill were appointed to the commissariat at Hong Kong, where also the weakly men were left under command of Captain Raymond and Ensign Rennick, and attached to the provisional depôt battalion there formed. 1860

On the 16th of June the regiment arrived at Talienwhan Bay, near the entrance of the Gulf of Pecheli, where, eventually, the troops were landed and employed in digging wells. The expedition re-embarked on the 24th of July, and sailed for the Peiho River, near the mouth of which it anchored four days afterwards, but moved to the Pehtang River on the 30th. Here the regiment disembarked* on the 6th of August, and on the 12th the troops advanced to attack the Tartar posts at the Sin-Ho intrenchments. This delay was occasioned by the very

* The strength of the regiment landing at Pehtang was twenty-seven officers and nine hundred and thirteen men of all ranks. The officers were :—

Lieut.-Colonel MacMahon.
Brevet Lt.-Colonel Browne.
Major Hackett.
Brevet Major Robinson.
" " Preston.
Captain Gregory.
" Walters.
" Ingham.
" Mansergh.
" Bower.
Lieutenant Wood.
" Smith.
" Pigott (Adjt.)
" Birch.

Lieutenant Rogers.
" Hodson.
" Pitt.
" Foley.
" Heane.
" Rennie.
Ensign Roberts.
Surgeon Mee.
Assistant-Surgeon Kinahan.
" " McDowell.
" " Orton.
Paymaster Thomsett.
Quartermaster Hart.

1860 heavy rains, which had made the country nearly impassable. The ground upon which they were about to advance was at any time of a most difficult nature, and intersected with broad and deep canals used by the Chinese for the manufacture of salt. The 44th formed a portion of the second division, under Major-General Sir Robert Napier, whose force commenced to move out of their quarters at four o'clock in the morning of the 12th of August, in the following order:—An advanced guard of two hundred men of the 3rd Buffs, with two Armstrong guns of Milward's battery, under the command of Lieut.-Colonel Sargent, of the 3rd Foot; four Armstrong guns of the same battery, the 23rd Company Royal Engineers, 3rd Buffs, 8th Punjaub Infantry, 44th Regiment, Rotton's rocket battery, Royal Marines, Madras Sappers and Miners, right wing of the 67th, reserve ammunition, hospital stretchers, &c. Rear guard—left wing of the 67th Regiment; the cavalry brigade, comprising two squadrons of the 1st or King's Dragoon Guards, Probyn's Horse, three guns, Stirling's battery, and Fane's Horse.*

Although two days' hard work had been devoted to the repairing of the roads, yet the deep tenacious mud rendered them so difficult, that the rear of the column did not clear the gate of Pehtang until half-past seven o'clock. It cost the troops two hours of hard labour to traverse the first two miles.

The plan of operations was for Sir John Michel's division to move straight along the causeway, and carry the intrenchments defending it, while Sir Robert Napier made a detour with his division to the right, taking the enemy's position in flank.

* The main column, under Major-General Sir John Michel, which moved along the causeway leading directly from Pehtang to Sin-Ho, was preceded by the first infantry brigade, a company of Royal Engineers, an Armstrong battery, one thousand French infantry, and a French battery, the whole forming an advanced guard, under the command of Brigadier Staveley, of the 44th Regiment, and which carried the intrenchments at the end of the causeway.

Sir Robert Napier perceiving the enemy in great 1860
force both in the intrenchments and in front of the village of Sin-Ho, after gaining his flank, marched directly towards him, threatening also his line of retreat. On arriving within fifteen hundred yards, Milward's Armstrong guns opened; these were the first shots fired with that weapon in actual warfare, and their range and accuracy excited great admiration. This fire surprised the Tartar horsemen, but did not shake them; after some hesitation they poured out in a long line through a passage across the marsh which separated them from the British, and forming with great regularity and quickness, enveloped Sir Robert Napier's force in a great cordon of skirmishers. An opportunity was afforded for the cavalry to charge, and although the Chinese eventually gave way and fled with precipitation from the field, they evinced considerable personal courage under a heavy fire of artillery.

No opportunity was afforded throughout the day of coming in contact with the Chinese infantry. A cloud of Tartar cavalry, skirmishing, threatened the artillery of the brigade, but were driven off by four companies of the 44th wheeled up and firing volleys. Captain Bower, with the rear-guard, in charge of ammunition, also received and repulsed a charge of Tartar cavalry. Only four men of the regiment were wounded. The British casualties were limited to one man killed, and four officers and twenty-three men wounded. These included those which were incurred in the attack of the intrenched fortified camp of Tangku on the 14th of August. This small loss was partly attributable to the enemy being paralyzed by the fire of the artillery. On the latter occasion the regiment was in reserve.

After the capture of the fortified town of Tangku, the next place taken, Sir James Hope Grant commenced bringing up siege guns and ammunition from Pehtang with a view to the reduction of the principal fort on the left bank and near the mouth of the Peiho, about two miles distant from Tangku. On the 19th of August the

1860 regiment moved from Tangku to an encampment near the North Taku Forts.

Major-General Sir Robert Napier was placed in charge of the advance, his division being quartered at Tangku. By the 20th of August the road was made practicable to within eight hundred yards of the fort, batteries were traced, and the heavy guns were brought out in readiness to be placed in position by daybreak of the following day, at which time every thing was in readiness for the attack. The regiment left the encampment this day for a position near the fort selected for attack, and sent out strong working parties during the night to prepare batteries for the artillery. So eager were the Chinese for the fight, that they opened fire upon the troops at five o'clock on the morning of the 21st, from all their forts within range, causing the allied forces to commence an hour earlier than had been arranged

The storming party of Infantry consisted of a wing of the 44th, under Lieut.-Colonel Patrick William MacMahon, and a wing of the 67th, under Lieut.-Colonel Thomas, supported by the other wings of these two regiments, and the Royal Marines, under Lieut.-Colonel Gascoigne. A detachment of the latter, under Lieut.-Colonel Travers, carried a pontoon bridge for crossing the wet ditches, and Major Graham, of the Royal Engineers, conducted the assault, the whole being commanded by Brigadier-General Reeves.

About seven o'clock the enemy's magazine blew up with a terrific explosion, and a few minutes later the one in the outer North Fort was also exploded by a shell from the gunboats.

The firing of the forts having almost ceased, a breach was commenced near the gate, and a portion of the storming party advanced to within thirty yards to open a musketry fire—the French Infantry being on the right and the British on the left. This advance caused the allies partially to slacken the fire of their artillery, when the enemy, emerging from their cover, opened a heavy musketry fire upon the advancing troops. So vigorous 1860 was the resistance of the Chinese that the French, after having crossed the wet ditches in the most gallant manner, were unable to escalade the walls. The efforts of the Sappers to lay down the pontoon bridge were unavailing, no less than fifteen of the men carrying it being knocked over in one instant, and one of the pontoons destroyed. The difficulties were therefore considerable, and the troops had to wade through deep mud and swim ditches, the banks of which were thickly planted with sharp stakes.

Sir Robert Napier at this crisis had two howitzers of Captain Govan's battery brought up to within fifty yards of the gate, with the view more speedily to create a breach. A space sufficient to admit one man was soon made, and the storming party forced their way in by single file in the most gallant manner.*

Lieutenant Robert Montresser Rogers, commanding the leading (letter E) company, Private John M'Dougall, of the 44th, and Lieutenant Edmund Henry Lenon, of the 67th Regiment, after swimming the ditches, entered the North Taku Fort by one of the apertures during the assault, and were the first of the British established on its walls. They passed through in the same order as their names are recorded, each assisting the other to mount the embrasure, which was climbed by sticking bayonets into the wall. This gained for them the much prized Victoria Cross, which was also conferred upon Lieutenant Nathaniel Burslem, Ensign John Worthy Chaplin, and Private Thomas Lane, of the 67th Regiment, for similar gallantry. At the same time the French had effected their entrance by escalade, and the garrison, driven back step by step, were hurled pell-mell through the embrasures on the

---

* Brigadier Reeves, who commanded the troops for the assault, was severely wounded in five places, but did not quit the field until the success of the assault had been assured. Lieut.-Colonel MacMahon, 44th, however took command of the brigade, having previously led the storming party of the Regiment, and Brevet Lieut.-Colonel Browne succeeded to that of the 44th, and brought the regiment out of action.

1860 opposite side, where the same obstacles which had formed an impediment to the advance of the allies proved an obstruction to the retreat of the Chinese. In addition to two wet ditches and two belts of pointed bamboo stakes, there was, besides swampy ground, a third ditch and bank.

A destructive fire was opened upon the enemy by the storming parties from the Cavalier, enhanced by the canister fire of Captain Govan's guns, which had been moved to the left of the fort to bear on them, causing the ground outside of it to be covered with the dead and wounded of the garrison. About an hour afterwards the whole of the forts on both sides of the river hoisted flags of truce. When summoned to surrender, however, an evasive and insolent reply was given, the allies being defied to advance to the attack. After evincing considerable obstinacy, the Tartars were compelled to succumb, and the allied Infantry, pushing on towards the outer North Fort, scaled the walls without further opposition, and made prisoners the garrison of two thousand men. Towards evening the Chinese were seen evacuating the South Forts. Detachments of British and French were therefore passed over in ships' boats to occupy them.

This success was not gained without severe loss. The 44th had Captain George Ingham and Lieutenant Robert Montressor Rogers* severely wounded, fourteen men killed, one drummer and forty-five men wounded.

The words "TAKU FORT" were authorised by Her Majesty to be borne on the colour of the regiment, to commemorate its gallantry at their capture.

Shanghai being threatened by the Taeping rebels, the 44th Regiment was despatched at a few hours' notice, on

* Lieutenant Rogers was wounded by a musket ball some time before entering the fort. He was the first in of any officer or man of the English Army. In company with Private M'Dougall and a few of the French he afterwards rushed into the interior of the fort. Seeing the colours of the 67th Regiment, he called to the officer carrying them, went up, and stood by the side of Ensign Chaplin while they were placed on the ramparts.

the 25th August, for its protection, landed there on the 1860
10th of September, and was quartered in Joss-houses and Yamuns, in the Chinese city. The war terminated on the 13th of October by the allies jointly occupying one of the gates at Pekin, the ratification of the former treaty, and the payment of a large sum of money by the Chinese Government.

The 44th continued at Shanghai, where the regiment suffered much from fever and ague, consequent on the wretched accommodation and unhealthiness of climate, until the 15th of November, when it was removed to Hong Kong, landing there on the 27th November. From this date the command of it devolved upon Brevet Lieutenant-Colonel Andrew Browne, Lieut.-Colonel MacMahon having been appointed commandant at that island. Both these officers, for their services in China, were made Companions of the Order of the Bath. Captain Gregory † and the Honourable R. Baillie Hamilton were promoted to the rank of Brevet-Major, and Lieutenant Rogers to an unattached company, the latter in addition to the award of the Victoria Cross.

Upon Brevet Lieut.-Colonel Browne returning to England, in February, 1861, the command of the regiment was assumed by Major John Hackett, Brigadier Staveley, who received the distinguished service pension, having continued in command of the force at Tien-Tsin, with the rank of Brigadier-General.* Whilst stationed at Hong Kong the regiment was divided into wings, one being

† Captain Gregory was one of the first into the Taku Fort after those who obtained the Victoria Cross. The conduct of Lieutenant Birch was specially noticed; also that of Assistant-Surgeons Kinahan and M'Dowell, of the regiment, who displayed great zeal and exertion in bringing up stretchers and in attending to the wounded on the field under the heaviest fire.

* Colonel Staveley subsequently succeeded to the command of the troops in China, and commanded the French and British Forces when clearing out the Taepings from a radius of thirty miles around Shanghai.

1861 quartered there and the other in huts and tents at Kowloon, on the main land opposite.

The summer of 1861 at Hong Kong was an unusually healthy season, and the regiment had but little sickness for a quarter so notoriously unhealthy.

In September the 44th received the order for its return to India, and, on leaving China, the following order and letter were published to the regiment:—

" *Head-quarters, Hong Kong,*
" *13th October, 1861.*

" *Division Orders.*

" No. 1.

" H.M.'s 44th Regiment being about to leave this " command, Brigadier-General Crawford, C.B., command- " ing South China, begs to convey to Colonel MacMahon, " C.B., the commandant, and to Major Hackett, command- " ing the regiment, his unqualified approbation of the state " of the regiment.

" The Brigadier-General has been unable to make the " half-yearly inspection, in consequence of the continued " heat of summer and his own protracted illness, but, from " his constant and close observation, he is perfectly satisfied " that the regiment leaves this command in the highest " state of discipline and efficiency.

" The interior arrangement of the 44th Regiment " appears to be admirable, ensuring correctness and preci- " sion in all that is required from a regiment.

" To officers, non-commissioned officers, and men of " H.M.'s 44th Regiment, the Brigadier-General tenders " his best thanks for the admirable conduct of the regiment " whilst under his command, and best wishes for their " prosperity.

" By order,
" (Signed) C. F. Grant, *Capt., D.A.A.G.*"

" *Head-quarters, Hong Kong,* 1861
" *Oct. 13th,* 1861.

" No. 503.

" Sir,—The Brigadier-General Commanding in South " China has forwarded to Sir J. Michel his Order No. 1 of " this date.

" In every word of that Order the Major-General " desires me to say he heartily concurs, and directs that " you will notify to the regiment his approbation of their " great merit as a corps.

" He will not fail to bring very prominently to the " notice of the Commander-in-Chief in Bombay, and also " of His Royal Highness the Duke of Cambridge, the " Order issued by Brigadier-General Crawford, and his " own thorough concurrence in its spirit.

" The Major-General much regrets that he was not " enabled to see the regiment before their embarkation, " and express to them in person his sense of the loss he " sustains by their departure.

" I have, &c.,

" (Signed) L. Mansergh, *Capt., D.A A.G.*

" *The Officer Commanding*
" *44th Regiment.*"

On the 11th, 12th, and 14th of October the 44th, being relieved by the 99th, from Canton, embarked for Bombay in three detachments, and on arrival at Vingorla, on the Malabar coast, during the month of December, proceeded to Belgaum, to relieve the 83rd, under orders for England.

The head-quarters, under Lieut.-Colonel MacMahon, arrived at Belgaum on the 4th of January, the other two portions of the regiment having reached that station on the 18th and 25th of the previous month. The women and children arrived here from Madras at the same time, most of them having been separated from their husbands and fathers for upwards of four years.

1862 On the 2nd of October, 1862, at a garrison parade, the Victoria Cross was presented, by Brigadier Adams, C.B., to Private John M'Dougall, for his gallantry in the assault on the Taku Forts in China, as previously described. During this year the companies of the regiment were assimilated, the flank companies being dressed and equipped in the same manner as the remainder—the Light company became letter "I" company, and the Grenadier company, letter "H" company.

1863 Seven hundred and fifty-two medals for service in China were issued to the regiment on the 8th of January, 1863, by Brigadier Adams, C.B., commanding the Belgaum district. The remarks of His Royal Highness the Field Marshal Commanding-in-Chief, on the confidential reports of the various inspecting officers, bear witness to the satisfactory state of the regiment at this period; the absence of crime was particularly remarked on, and it is also worthy of notice that there were no less than 502 depositors in the Regimental Savings' Bank.

1865 The most conspicuous event in the year was the Industrial Exhibition of articles manufactured in the workshops, or by the school industrial class: this Exhibition was the second of the kind held in the army. In this year Major-General Reed (Colonel), Colonel Staveley, and Major-General the Hon. A. A. Spencer were appointed K.C.B's. The regiment was again commended this year for its high state of discipline, and internal economy. The gardens allotted to the regiment appear to have been in a most flourishing condition at this period; the value of the surplus produce alone amounting to over 1,200 rupees.

Towards the end of the year the regiment being under orders for England, gave 287 volunteers to other corps, and on the 20th of November, 3 companies under Captain Linton, and Lieutenant Sutherland embarked on board the transport "Tweed"; followed on the 23rd of December, by the head quarters and remainder of the regiment, in the "Dilawar."

Shortly before the embarkation, the following very complimentary General Order, dated Poona, 16th December, 1865, was received from Sir Robert Napier, K.C.B., Commander-in-Chief of the Bombay Army:— 1865

"The Commander-in-Chief regrets that he cannot "have an opportunity of seeing the 44th regiment before it "passes from his command. Sir Robert Napier was "associated with the regiment during the campaign in "China, where it was particularly distinguished for its "gallantry in the assault of the North Taku Forts. The "report of the high state of discipline of the corps during "its service in this presidency, has given the Commander-"in-Chief great pleasure, and in conveying his best wishes "for the future welfare of Colonel MacMahon, and the "officers and men of the 44th regiment, His Excellency "feels assured that they will maintain their discipline and "efficiency which marked their career here, and which will "ensure their success wherever their services may be "called forth.

"By order of His Excellency, the Commander-in-Chief "—Copy of this order sent to the Adjutant-General for "publication; forwarded direct to Colonel MacMahon, "C.B., Commanding 44th Regiment, by desire of His "Excellency Sir Robert Napier, K.C.B., to ensure its "receipt before embarkation of the regiment.

"By command,
" W. A. DILLON, Lt.-Col.
" Military Secretary."

The following officers embarked with the regiment:—

Col. P. W. MacMahon, C.B.
Captain F. D. Walters.
,, E. A. Raymond.
,, G. C. Bower.
,, W. A. Wood.
,, M. S. Smith.
Lieut. C. Maguire.
,, K. Browne.

Lieut. W. H. Overton, *Acting Adjutant.*
,, G. Bain.
Ensign H. Gordon.
,, W. Wood.
,, H. S. Gordon.
Surgeon D. Macqueen.

1866 During the voyage home in the "Dilawar," on the 15th of March, 1866, a fire broke out in the spirit room, which was extinguished with the greatest difficulty, as there was a strong wind, and the ship was rolling heavily.

The conduct of the regiment on this most trying occasion was admirable, and the following regimental order was issued the day after:—

"*Transport* "*Dilawar*,"
"*At Sea, March* 16*th*, 1866."

"In giving expression to the deep thankfulness which "must be felt by all, for our merciful and providential "escape of yesterday, by the fire which broke out on board "having been so promptly extinguished Colonel "MacMahon desires to record the high approval and "satisfaction which the soldierlike and orderly conduct of "the regiment elicited on this occasion. The steadiness "and active exertions of all ranks were most creditable to "the corps, and deserving of all praise; while all "admirably performed their allotted duty, many were most "conspicuous for the intelligent assistance which they "afforded in supplementing the exertions of the officers "and men of the ship."

The regiment landed in England on the 28th of March, 1866, and proceeded to Dover, where it was joined by the detachment embarked in the "Tweed," and the depôt from Colchester. The establishment was fixed at 4 Field Officers, 10 Captains, 12 Lieutenants, 8 Ensigns, 5 Staff, 48 Sergeants, 21 Drummers, 40 Corporals, 560 Privates; the whole divided into 10 companies.

1867 The regiment left Dover on the 8th of March, 1867, and proceeded by rail to Aldershot, where it was attached to the 2nd Infantry Brigade, and quartered in the West Block Permanent Barracks. Snider rifles were issued to the regiment in January of this year.

1868 The regiment left Aldershot on the 22nd of January, 1868, and embarked in the "Simoom" for Kingstown. The head-quarters and 4 companies proceeded to Kilkenny, 4 companies, under Lieut.-Colonel Browne, to Templemore, and 2 companies, under Major Geddes, to Birr; and in April two of the companies stationed at Templemore were detached to Tipperary under Captain Linton. 1868

Gold lace accoutrements and trouser stripes were ordered to be worn this year by all officers of Infantry when in full dress.

During the month of June, 1869, the Glengarry forage cap was taken into wear by the regiment. On the 27th of July the regiment moved to the Curragh, and thence, on the 22nd of October, to Inniskillen, furnishing detachments of 4 companies at Newry, 2 companies at Armagh, and 1 company at Monaghan. 1869

In July, 1870, the regiment was concentrated at the Curragh, in view of expected hostilities at the Cape of Good Hope, and the strength was raised to 800 rank and file: its services however were not required. 1870

Private George Betts, of "G" company, was unfortunately drowned at Belturbet, in June, in spite of a gallant attempt to rescue him, made by Lieutenant R. B. Bald, who succeeded by diving, in recovering the body and bringing it to shore, but too late to restore animation. Lieutenant Bald received the thanks of the Commander-in-Chief in Ireland, in a letter, stating "His Lordship has received the report of the promptitude and gallantry displayed by Lieutenant Bald on the occasion in question with the utmost pleasure and gratification."

The regiment was not destined to remain long at home, for orders for India were received on the 27th of July, 1871, the establishment at the same time being fixed at 1032 of all ranks. The regiment embarked at Queenstown in H.M.S. "Malabar" on the 27th of September, 1871, and having arrived at Bombay on the 1st of November, proceeded by rail to Kampti under the command of 1871

1871 Colonel Thomas Raikes, who had exchanged from half pay with Colonel Andrew Browne, C.B., and assumed command on the day the regiment embarked. The depôt, consisting of "C" and "G" companies and commanded by Captain Rennie, was stationed at Chatham on the embarkation of the regiment for India.

1872 The following is a list of the officers on the 1st of January, 1872:—

Lieut.-Col. T. Raikes, C.B. (Col.), Head Quarters, Kampti.
Major R. Preston (Lieut.-Col.), Head Quarters, Kampti.
Major F. D. Walters.
Captain W. A. Wood, Head Quarters, Kampti.
" C. E. Rennie, Depôt, Chatham.
" C. Maguire, Head Quarters, Kampti.
" W. J. E. G. Sutherland, " "
" E. C. P. Pigott, " "
" R. B. Bald, Depôt, Chatham.
" R. R. O'Grady, Head Quarters, Kampti.
" B. H. Foster, " "
" H. G. Lefroy, " "
" W. D. Saunders, " "
Lieut. C. H. Walsh, Head Quarters, Kampti.
" T. T. Irvine, " "
" W. Wood, I. of M., " "
" W. Odell, " "
" P. Edgcumbe, Depôt, Chatham.
" S. Churchill, Head Quarters, Kampti.
" H. F. Hill, " "
" T. S. W. Bernard, Depôt, Chatham.
" E. Morrell, Head Quarters, Kampti.
" H. A. Richards, " "
" A. J. Wilkinson, Leave to England.
" E. W. Davies, " "
Ensign G. H. Powell, Head Quarters, Kampti.
" M. Wynyard, " "
" E. N. Leahy, " "
" W. P. Kennedy, " "

Ensign A. G. F. Browne, Leave to England. 1872
„ C. D. Rosser, Head Quarters, Kampti.
Paymaster R. B. Farwell (Capt.), „ „
Adjutant A. J. Roberts, „ „
Quartermaster W. H. McHarg, „ „
Surg.-Major D. Macqueen, M.D., „ „
Asst.-Surg. A. H. Stokes, M.B., „ „
„ T. M. Kirkwood, „ „

The regiment remained at Kampti during the year 1872, the only event worthy of notice being the inspection by Sir Fred. Haines, K.C.B., Commander-in-Chief in Madras, on the 27th of November, and the visit of the Viceroy, and Governor-General, Lord Northbrook, on the 30th of the same month. His Lordship visited the barracks and afterwards inspected the European hospital.

The second Industrial Exhibition took place on the 27th of February, 1873, being opened by Colonel Raikes, C.B. This exhibition which was composed almost entirely of articles contributed by the artificers, and women and children of the regiment was most favourably noticed by the Commander-in-Chief of the Presidency. 1873

On the 12th of September, Quartermaster W. H. McHarg embarked for England, after a total service of 30 years with the regiment, 17 years of which had been as Quartermaster.

Authority was received during the year 1874 for the officers of the regiment to wear the Sphinx, with the word "Egypt" on the forage cap, above the numerals. 1874

The regiment left Kampti for Secunderabad on the 26th of November, 1875, and arrived at the latter station on the 2nd of December; the only other event of any importance about this time being the appointment of Lieutenant-Colonel Preston as Military Secretary to the Commander-in-Chief in India. 1875

The regiment took part in the full dress parade held on the 1st of January, 1877, to celebrate the Proclamation 1877

1877 of Her Majesty as Empress of India, and was much praised for its general appearance.

In June, 1877, the Martini-Henry rifle was issued in place of the Snider, and towards the end of the year the regiment proceeded to Burmah, in relief of the 89th Regiment, the right wing being stationed at Thayetmyo, and the left at Toungoo.

1879 The regiment remained in Burmah during the whole of the year, which, owing to the disturbances in Upper Burmah, appeared likely to be an eventful and stirring one. On the 10th of March telegraphic orders were received for the regiment to be held in readiness to proceed on service at a moment's notice, and all necessary preparations were at once made. The frontier garrison of Thayetmyo was at the same time reinforced by the 10th Battery, 8th Brigade, Royal Artillery, 43rd Light Infantry, a wing of the 89th Regiment, the 31st Madras Light Infantry, a wing of the 19th Madras Native Infantry, and a company of Sappers and Miners; making a total of 1555 Europeans, and 1629 Natives. Owing, however, to the outbreak of the second Afghan War, and other political reasons, no advance was made beyond the frontier.

An outbreak of cholera occurred during the early part of this year, and though every precaution was taken to keep the sickness down, by the formation of camps, and by starting a variety of games and amusements, a considerable number of deaths occurred. Shortly before the departure of the regiment from Thayetmyo, a brass tablet was placed in the Garrison Church, in memory of those men who died in Burmah, on this and subsequent occasions.

1880 In March, 1880, on the suggestion of Colonel Hand, the left wing at Toungoo proceeded across the Pegu Yoma Range, to join head-quarters at Thayetmyo. No difficulty was experienced in the march, which had never before been performed by European troops, except from the hilly nature of the track. A few weeks later "A" and "B" companies, under command of Captain Hill, crossed

the Irrawaddy to Tayrangoon, and together with two 1880
companies of the 43rd Light Infantry, relieved the detachments of the 89th Regiment stationed there.

It is a fact worthy of note that of the lengthened 1881
period of nearly 60 years, from the embarkation of the regiment for India in 1822 up to the year 1881, only ten years were passed at home; and during this time the regiment fought in Burmah, Affghanistan and China, and added to its former glory by its gallantry and fortitude during the protracted siege of Sebastopol.

A heavy loss was in store for the regiment this year, for on the introduction of the territorial organisation scheme on the 1st of July, 1881, the 44th became the 1st Battalion of the "Essex Regiment," the 2nd Battalion being composed of the 56th Regiment, while the Essex Rifles and the West Essex Militia formed the 3rd and 4th Battalions respectively.

And so passed away from the regiment the name under which it had served during well nigh a century and a half of arduous service—the name under which it had gained for itself undying renown alike "in the most brilliant successes, and in the direst disasters." But the 44th Regiment has long since learned, through terrible ordeals, to lightly regard even far greater hardships than the loss of a much loved and time-honoured title;—the glory remains—and though deeply regretting the hard fate that has rendered necessary the surrender of such a title, the regiment may look forward with hope to the day when, aided by its second battalion, it shall have raised up round the word "Essex" a halo of glory which will equal—for it could hardly suppass — that which surrounds the old two fours.

APPENDIX.

Succession of Colonels of the 44th Regiment.

Name.	*Date of Appointment.*
Colonel James Long	7th January, 1741.
Lieut.-Colonel John Lee..................	11th March, 1743.
Colonel Sir Peter Halkett, Bart.	26th February, 1751.
Colonel Robert Ellison	13th November, 1755.
Major-General James Abercromby ...	13th March, 1756.
Major-General Charles Rainsford ...	4th May, 1781.
General Sir Thomas Trigge, K.B. ...	27th May, 1809.
General The Earl of Suffolk	12th January, 1814.
Lieut.-General Gore Browne	28th January, 1820.
Lieut.-General The Honble. Patrick Stuart	7th April, 1843.
Lieutenant-General Sir Frederick Ashworth	8th February, 1855.
Major-General Sir Thomas Reed, K.C.B.	2nd August, 1858.
General Sir Charles Staveley, G.C.B.	25th July, 1883.

ccession of Commanding Officers of the Forty-Fourth from the Formation of the Regiment to the Year 1881.

Names.	*From*	*To*	*Remarks.*
Peter Halkett, art.	1 Feb., 1741..	26 Feb., 1751	Appointed Colonel of the 44th; killed in action near Fort du Quesne, 9 July, 1755.
ı. Thomas Gage..	2 March, 1751	5 May, 1758...	Appointed Colonel of the 80th (since disbanded); General 20 Nov., 1782; died 2 April, 1787.
liam Farquhar...	19 Sept., 1758	28 Oct., 1759.	Died at Oswego, in 1759.
liam Eyre.........	29 Oct., 1759.	11 Dec., 1764.	Died in 1764.
ıes Agnew.........	12 Dec., 1764	4 Oct., 1777...	Brigadier-General Agnew, killed in action at Germantown 4 October, 1777.
ıry Hope	5 Oct., 1777...	13 April, 1789	Brigadier-General Hope, Lieut.-Governor of Quebec, died 13 April, 1789.
liam John Darby	13 June, 1789	26 Nov., 1793.	Retired in 1793.
ɔert Riddell	26 Nov., 1793	27 Sept., 1796	Brigadier-General Riddell, died at St. Lucia 27 September, 1796.
vid Ogilvie.........	1 Sept., 1795.	4 April, 1801.	Died of wounds 4 April, 1801.
ırles Baillie	18 Jan., 1797.	25 July, 1798.	Exchanged to 50th Foot, and appointed to 51st Regiment; Colonel 1 January, 1800; Colonel of the 3rd Ceylon Regiment 7 April, 1804; Major-General 25 April, 1806; died in 1810.
ıncis Erskine......	25 July, 1798.	24 Jan., 1799.	Exchanged from 50th Foot, and afterwards to half-pay 99th Regiment: died 7 June, 1833.

Names.	*From*	*To*	*Remarks.*
Christopher Tilson...	24 Jan., 1799.	17 Feb., 1814.	Exchanged from half-99th Regt.; name chan to Chowne in Janu 1812; Lieut.-General June, 1813; Colonel of 76th 17 February, 18 General 22 July, 18 died 15 July, 1834.
Kenneth Mackenzie.	26 May, 1801, ante-dated 5 April, 1801.	24 Feb., 1803.	Promoted from the 90 exchanged to 52nd R ment; Lieut.-General July, 1821; created baronet and name chan to Douglas in 1831; Kenneth Douglas, Ba died 22 November, 183
Alexander Dirom ...	24 Feb., 1803.	4 June, 1814..	Exchanged from 52n Lieut.-Gen. 4 June, 18 died 6 October, 1830.
Thomas Nicoll	8 Oct., 1803...	15 June, 1804	Retired 15 June, 1804.
Arthur Brooke, C.B.	15 June, 1804	12 Aug., 1819	Lieut.-General 10 Janu 1837; Sir Arthur Bro K.C.B., died 26 July, 1
Robert Garden	1 Dec., 1805..	5 Jan., 1809...	Exchanged to 89th; Col 25 April 1808; died June, 1810.
Charles Bulkeley Egerton.	5 Jan., 1809...	22 Aug., 1811	Exchanged from 89th, to Inspecting Field Off of Militia in Canada 1811; General 9 Novem 1846; Sir Charles Eger G.C.M.G., and K.C died 8 July, 1857.
Hon. George Carleton	22 Aug., 1811	8 March, 1814	Exchanged from Inspec Field Officer of Militi Canada; killed in ac before Bergen-op-Zoo March, 1814.
John Millet Hamerton, C.B.	31 Mar., 1814	25 Mar., 1816	Placed on half-pay on bandment of the sec battalion; General 20 J 1854; died 28 Janu 1855.

Names.	From	To	Remarks.
George Henry 'edk. Berkeley, C.B.	12 Aug., 1819	22 Feb., 1821.	Exchanged to 3rd Foot Guards; General 20 June, 1854; died 25 September, 1857.
'ge Thomas ıpier, C.B.	22 Feb., 1821.	19 April, 1821	Exchanged from 3rd Foot Guards, and in 1821 to half-pay, Sicilian Regt.; General 20 June, 1854; Sir George Napier, K.C.B., died 8 September, 1855.
Wanton Morri-ı, C.B.	19 April, 1821	15 Feb., 1826	Exchanged from half-pay, Sicilian Regiment; Brevet-Colonel 12 August, 1819; Brigadier-General Morrison, died on passage home from India 15 February, 1826.
ge Hardinge ...	25 May, 1822.	25 Mar., 1824	Appointed to 96th Regiment 25 March, 1824; died 9 March, 1829.
Henry Dunkin, B.	25 Mar., 1824	11 Nov., 1825	Brevet-Colonel 12 August, 1819; died at Dacca 11 November, 1825.
ıcis Skelly Tidy, B.	4 May, 1826.. ante-dated 11 Nov., 1825	25 Aug., 1829	Promoted from 14th Foot; Inspecting Field Officer of a Recruiting District in 1829; Brevet-Colonel 22 July, 1830; appointed to 24th Foot 1 March, 1833; died whilst in command of the regiment at Kingston, Canada, 9 October, 1835.
ıilton L. Carter.	12 Nov., 1825	5 Sept., 1827.	Died at Ghazeepore 5 September, 1827.
Shelton	16 Sept., 1827	13 May, 1845.	Brevet-Colonel, 23 November, 1841; died 13 May, 1845.
rt Macdonald, B.	25 Aug., 1829	11 June, 1830	Promoted from 35th; exchanged to half-pay 12th Foot 11th June, 1830; appointed Lieut.-Colonel 35th 22 July, 1830; retired 12 April, 1831; died 15 November, 1860.

Names.	*From*	*To*	*Remarks.*
Thomas Mackrell ...	25 June, 1830	10 Nov., 1841	Died of wounds receive the storming of Rika-hee Fort, near Cabul November, 1841.
Hon. Augustus Almoric Spencer, C.B.	17 May, 1845.	1 Aug., 1856.	Placed on half-pay of 44th 1st August, 1
James Oliphant Clunie, C.B.	25 Dec., 1847.	24 Nov., 1848	Retired 24 November, 1 died 27 July, 1851.
Augustus Halifax Ferryman, C.B.	24 Nov., 1848	4 Sept., 1849.	Exchanged to 89th in 1 Colonel 28th Novem 1854.
Edward Thorp	4 Sept., 1849.	27 June, 1851	Exchanged from 89th; pointed to 21st Foot o June, 1851; retired April, 1852; died 25 J ary, 1853.
Sir Charles William Dunbar Staveley, K.C.B.	9 March, 1855	8 March, 1865	Colonel 9 March, 1858.
Patrick William Mac Mahon, C.B.	28 Aug., 1857	28 Dec., 1866	Colonel 4 May, 1861.
Col. J. J. Hort	28 Dec., 1866.	10 Nov., 1869.	
Col. A. Browne, C.B.	10 Nov., 1869.	27 Sep., 1871.	
Col. T. Raikes, C.B.	27 Sep., 1871.	11 Feb., 1875.	
Col. R. Preston	11 Feb., 1875.	27 Oct., 1876.	Appointed Military Se tary to the Comman in-Chief in India.
Col. J. S. Hand	27 Oct., 1876.	———	

Names of Men who Served in the Trenches before Sebastopol the Whole Time.

Number.	*Rank.*	*Names.*	*Company.*	*Regimental Number.*	*Rank.*	*Names.*	*Company.*
9	Sergt.-Major	David Watson	—	1,702	Private	Patrick Walsh	5
1	Q.M. Sergt.	Denis Reddin*	—	2,504	Colour-Sergt.	Phillip Holland	6
8	Sergeant	Alexander McGregor	1	2,515	Sergeant	John Bartley	6
1	Private	Michael Bannon	1	2,634	"	Oscar Kelly	6
7	"	George Berryman	1	2,239	"	Patrick Burns	6
9	"	Denis Canty*	1	3,013	Private	John Burnside*	6
1	"	John Carroll	1	3,807	"	Jeremiah Donovan	6
0	"	Robert Crookshanks*	1	2,699	"	Patrick McGlew	6
8	"	Bartholomew Duff	1	2,492	"	Richard Mulholland	6
6	"	Michael Finn	1	3,252	"	Thomas Rigney	6
0	"	Robert Kerr	1	2,046	Sergeant	George Murray	7
8	Colour-Sergt.	John Milroy	2	1,440	Private	William R. Brown	7
6	Corporal	William Cullen	2	1,838	"	Charles Gravett	7
9	Private	James Kelly	2	1,881	"	Edwin Lingwood	7
7	"	Michael Lemasney	2	1,809	"	Nicholas McCarthy	7
2	Sergeant	William Trinder	3	1,988	"	Hugh Mearns	7
2	Corporal	John Drenon*	3	3,019	"	John Murley	7
0	Private	Henry Bird	3	2,773	"	Joseph Ward	7
2	Sergeant	Thomas Brown*	4	2,802	Sergeant	William McWhiney‡	8
5	Corporal	Robert Crook	4	1,874	"	Thomas Randall	8
8	"	Edward Hilditch	4	3,655	"	William Simpson	8
2	Private	John McCarter	4	3,576	Corporal	Patrick Fahey	8
7	Drummer	Thomas Durr	4	2,914	"	James McCarthy	8
7	Private	Charles Bibby	4	2,088	Private	Martin Brophy	8
5	"	George Carnt	4	1,650	"	Robert Crothers	8
4	"	Richard Dillon	4	3,670	"	George Finch	8
9	"	Thomas Irvine	4	2,436	"	Robert Graham	8
8	"	Michael Ledwigge	4	3,329	"	Thomas McCarthy*	8
1	"	Thomas McGuire	4	1,816	"	Joseph Mason	8
1	"	Joseph Sewell	4	1,821	"	William Massey	8
4	"	George Taplin	4	3,608	"	James Meades	8
1	Colour-Sergt.	Patrick Torpy*	5	3,204	"	Michael Neill	8
7	Sergeant	James Philips	5	3,407	"	William Newton	8
6	Corporal	John Shea	5	2,034	"	Robert Oakland	8
1	Private	Benjamin Conlan	5	3,824	"	Thomas Power	8
5	"	Andrew Dougherty	5	3,150	"	Thomas Rudge	8
4	"	Christopher Keefe	5	3,816	"	Michael Swift	8
0	"	James McGill	5	3,835	"	William Weston	8
0	"	Francis Mulholland	5	3,351	"	William Woodgate	8
6	"	Hugh Robinson	5	3,792	"	John Caharine	8
1	"	Lumsden Taylor	5				

* *Received special marks of distinction, vide page* 168.

‡ *Received the Victoria Cross, vide page* 169.

Names of Men who were in the Crimea the Whole Time.

Regimental Number.	*Rank.*	*Names.*	*Company.*	*Regimental Number.*	*Rank.*	*Names.*
2,064	Sergeant	William Powell	1	2,070	Corporal	Bernard Feeley......
2,903	,,	Abraham Friend ...	1	3,721	Private ..	Robert Kelly ....
2,195	,,	Henry Hall	1	3,060	,,	Samuel Shambrook..
3,604	Private ..	WilliamStarsemeare	1	3,205	,,	Edward Slaven ......
3,471	Sergeant	Henry Clarke.........	2	2,998	,,	Henry Warren ......
2,896	,,	Edmond Gill	2	1,786	Cr.-Srgt.	William Burrett ...
3,339	,,	Samuel Thompson...	2	3,255	Private ..	James Henry.........
3,701	Private ..	James Clinton	2	3,473	,,	Robert McDonald ...
2,233	,,	Henry Carroll*	2	2,885	,,	David Marforum ...
2,741	,,	Thomas Hull	2	2,350	,,	James Schofield......
2,095	,,	John Mealigh.........	2	2,221	Cr.-Srgt.	Thomas Mathews ...
3,642	,,	William Morris	2	1,788	,,	Peter Cantwell ......
3,452	,,	George Rouse	2	1,653	,,	Edward Kearns......
2,408	,,	William Broderick...	3	3,373	Corporal	Thomas Allen ......
1,419	,,	Job Mason	3	2,008	,,	James Milroy.........
3,134	,,	William Steele	3	2,001	,,	William Pegg.........
1,669	,,	James White	3	2,718	Private ..	William Drake ......
2,932	Sergeant	Isaac McElroy	4	3,177	,,	William Gallagher...
2,607	Private ..	George Saunders*...	4	2,425	,,	John Gallway ......
2,582	,,	Jonathan Storey ...	4	2,245	,,	Jacob Jenkinson ...
2,566	,,	William Tramain ...	4	3,666	,,	Phillip Lambert......
3,676	Drummr.	Joseph Patterson ...	5	1,826	,,	George Meades ......
3,653	Private ..	Michael Crowley. ...	5	3,515	,,	John Moore ... .
2,968	,,	Francis Kelly.........	5	3,293	,,	James Rankin ......
3 686	,,	John Roche............	5	1,515	,,	James Samples* ...

* *Received special marks of distinction, vide page* **169**.

Names of Men mentioned in Regimental Orders during the Crimean War.

Number.	Rank.	Names.	Regimental Number.	Rank.	Names.
99	Sergt.-Mjr.	David Watson.	2,871	Private ..	Benjamin Conlan.
04	Color-Srgt.	Phillip Holland.	3,807	,,	Jeremiah Donovan.
21	,,	Thomas Mathews.	3,670	,,	George Finch.
48	,,	John Milroy.	1,838	,,	Charles Gravett.
29	,,	Richard Rice.	3,830	Lce.-Corpl.	Robert Kerr.
34	Sergeant...	Oscar Kelly.	3,290	Private ...	James McGill.
418	,,	Alexandr. McGregor	2,662	,,	John Rowley.
46	,,	George Murray.	3,475	,,	Robert Thimbleby†
27	,,	James Phillips.	2,773	Lce.-Corpl.	Joseph Ward.
736	Lnce.-Srgt.	John Shea.	1,998	Sergeant ...	Robert Mann.
60	Private ...	Henry Bird.	2,696	Corporal ...	William Cullen.
408	,,	William Broderick.	3,351	Lce.-Corpl.	William Woodgate*

Names of Men who distinguished themselves during the Chinese Campaign of 1860.

Number.	Rank.	Names.	Regimental Number.	Rank.	Names.
220	Private ...	John McDougall.‡	7	Private ...	James Shields.
280	,,	Michael McCaffrey.	2,504	Color-Srgt.	Phillip Holland.
328	,,	James Bolger.	4,473	Drummer..	Stephen Lawson.
425	,,	Robert Alcock.	4,298	Private ...	Patrick Keough.
645	,,	Michael Driscoll.	2,939	,,	John Gallay.
193	,,	Joseph Fenton.	4,152	,,	John Daly.
614	,,	David Saunderson.	3,377	,,	Alex. McCann.
171	,,	William Duffey.	3,951	,,	Timothy Shea.
501	,,	Thomas Dyer.			

* *Received special marks of distinction, vide page* 168.
† *Received the insignia of the* 5*th class of the Legion of Honour, vide page* 168.
‡ *Received the Victoria Cross, vide page* 177.

www.ingramcontent.com/pod-product-compliance
Ingram Content Group UK Ltd.
Pitfield, Milton Keynes, MK11 3LW, UK
UKHW021835270726
14058UKWH00002B/163

9 781843 422440